HUMAN ANATOMY
MADE SIMPLE

HUMAN ANATOMY MADE SIMPLE

I. MacKay Murray, M.D.

Professor of Anatomy
State University of New York
Downstate Medical Center

Illustrated by Eva Cellini

MADE SIMPLE BOOKS
DOUBLEDAY & COMPANY, INC.
GARDEN CITY, NEW YORK

Library of Congress Catalog Card Number 68–22473
Copyright © 1969 by Doubleday & Company, Inc.
All Rights Reserved
Printed in the United States of America

ABOUT THIS BOOK

Anatomy is, broadly speaking, the analysis of a structure with reference to its parts; that is, the study of the parts that go to make something up. The structure can be any "thing": an airplane, a ball-point pen, a flower, or a chipmunk. It follows, then, that human anatomy is taken to mean the study of the structure and the organs of man.

To the extent that anatomy can be learned from studying a book, *Human Anatomy Made Simple* is ideal for self-teaching. The subject is presented in logical sequence, and each new term is explained when it is introduced. The drawings form an integral part of the book and must be studied along with the text. Many of the drawings are not exact anatomical replicas of the human body, but were designed to convey a particular idea in the simplest way possible, since it is understood that most readers of this book will not have access to models or to a body for dissection.

Another unique feature of this book is that it uses what is known as "living anatomy" to teach the subject. Living anatomy has advantages over studying anatomy from a dead body because structures can be observed as they actually function. By "taking your pulse" you actually feel the artery pulsate as the heart pumps blood. Similarly, you can test a muscle and feel it contract. By using your own body as much as possible you will gain the most from your study of human anatomy.

—THE PUBLISHER

CONTENTS

CHAPTER 7

CHAPTER 8

CHAPTER 9

HUMAN ANATOMY
MADE SIMPLE

CHAPTER 1

GENERAL TERMINOLOGY

THE ANATOMICAL POSITION

In order to understand the diagrams and descriptions of the human body, it is necessary to adopt a standard body position. The one universally used is the **anatomical position** (Fig. 1).

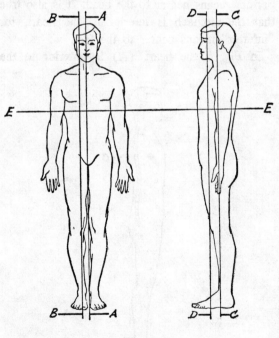

Fig. 1

The subject stands erect with toes pointing forward, eyes looking forward, arms by the side of the body with the palms of the hands facing forward. In future descriptions of the position of any structure it will be assumed that the body is in the anatomical position regardless of the actual position of the person.

The Planes of the Body. As a further aid it has been found useful to divide the body by imaginary planes. The **median plane** (shown as A in Fig. 1) is an imaginary vertical plane dividing the body into right and left halves. *Any* imaginary vertical plane parallel to the median plane is called a **sagittal plane** (B). The body may be divided by many sagittal planes but by only one median plane. The **frontal plane** (C) is *any* imaginary vertical plane at right angles to the median plane. The plane D is also a frontal plane. The **transverse plane** (E) is *any* horizontal plane through the body at right angles to both the median and frontal planes.

In the diagrams of the various parts of the body that will be shown later, it is frequently stated that the diagram is of a **sagittal section** or of a **horizontal section**. This means that the part of the body being discussed has been cut along either the sagittal or the horizontal plane and you are looking at the cut surface. In a sagittal section of the head, for example, the site of the cut to be made in the sagittal plane is shown in Fig. 2 at A. A drawing is then

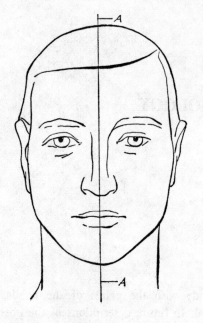

Fig. 2

made of the structures exposed on either of the cut surfaces, as in Fig. 3.

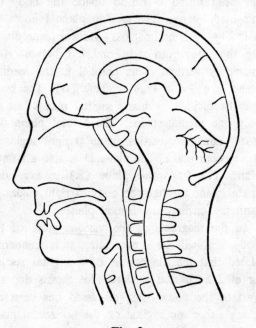

Fig. 3

Relative Position of One Structure to Another.
If we were to describe the relation of the heart

to the stomach in Fig. 4, we would say that the heart (A) is **superior** to the stomach (B). "Su-

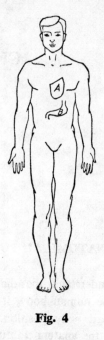

Fig. 4

perior" means nearer to the head. It is also true that the stomach is **inferior** to the heart, for "inferior" means nearer to the feet.

In Fig. 5 the heart (A) is **anterior** to the

Fig. 5

vertebral spine (C); "anterior" means nearer to the front or anterior surface of the body. Sometimes the term **ventral** is used instead of anterior. The vertebral spine lies **posterior** to the heart; "posterior" means nearer to the back. The term **dorsal** may be used instead of posterior.

The term **medial** means nearer to the median plane, and **lateral** means farther from the median plane. In Fig. 6 several medio-lateral relation-

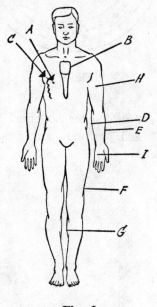

Fig. 6

ships are illustrated. The nipple (A) lies lateral to the sternum, or breast bone (B). However, the same nipple lies medial to the scar (C). In the anatomical position, the inside of the upper limb is called its medial side (D) because it is the side nearest the median plane. The side farthest from the median plane is the lateral side (E). G is the medial side and F the lateral side of the lower extremity.

We have previously used the term anterior or ventral. The nipple, for example, is on the ventral surface of the body. The front surface of the arm, indicated at H, is called its anterior or ventral surface. The opposite surface—that is, the back of the arm—is called the posterior or dorsal surface. Sometimes the ventral surface of

the hand (I) is called its **palmar** surface, or simply its palm.

Two terms frequently used in describing the relative position of structures in the limbs are **proximal** and **distal.** A reference point is always assumed; this is the root or attachment of the limb to the trunk. In Fig. 7 the root of the upper

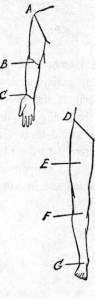

Fig. 7

limb is at A and that of the lower limb is at D. The elbow (B) is proximal to the wrist; "proximal" means nearer to the root of the limb, and "distal" means farther from the root. The fingers, for example, are distal to the wrist joint (C). Similarly, in the lower extremity the knee (F) is distal to the point E on the thigh. The knee is proximal to the ankle (G).

Superficial and **deep** are used to describe the relative position of structures with respect to the skin. Superficial means nearer to the skin, while deep means farther from it. For example, in Fig. 8, which is a transverse section through a hypothetical extremity, the skin (A) is shown covering its surface. C represents a blood vessel that lies deep to the nerve (B); or the nerve may be described as lying superficial to the blood vessel.

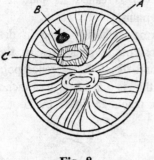

Fig. 8

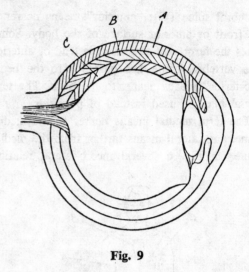

Fig. 9

External and **internal** are terms used when describing a body cavity, such as the abdomen, or an organ, such as the eyeball. In a section through the eyeball, Fig. 9, the sclera (A) is external to the choroid layer (B). Similarly, the retina (C) is internal to the choroid.

It is permissible to use **downward** and **below** for inferior, **forward** and **in front of** for anterior, **backward** and **behind** for posterior. Terms such as "under" and "on" are ambiguous and are best avoided.

CHAPTER 2

BONES AND JOINTS

Introduction. The bones of the human body make up about 15 percent of the body's total weight. They provide a rigid framework for the organs and tissues. The bones of the skull give protection to the fragile brain, while the bony ribs shield the heart and lungs from external forces. In the limbs, the bones provide rigid levers which are manipulated by the pull of the limb muscles to produce movement.

Bones consist essentially of tissue fibers which are impregnated with calcium salts. If a bone is soaked in dilute acid the calcium salts are removed, leaving only the tissue fibers. The "bone" is now very flexible and can be tied into a knot. If a bone is heated to a very high temperature the tissue fibers are destroyed, leaving only the calcium salts. Now the "bone" is very brittle. Children's bones, since they contain a higher proportion of fibers, are more likely to bend or crack, rather than break, from an external force. The bones of elderly people have lost tissue fibers as well as calcium salts and so are much more fragile.

A joint is that area where two bones are joined together. Movement cannot take place at all joints. Some bones are held together by **cartilage,** the so-called "gristle" of bone, which permits little movement. Other bones are held together by more flexible material, called **connective tissue,** which permits greater movement. Movable joints like the shoulder, hip, and knee receive the necessary support from thickened straps of connective tissue called **ligaments.** The muscles that pass over the joint also give added support.

BONES

Cartilage. A bone such as the **humerus,** or arm bone, first appears in the early embryo, but it consists entirely of cartilage, not bone. Cartilage is somewhat similar to a very firm gel, and it contains special cells that will be necessary later for the formation of bone. The shape or form of the cartilage resembles in miniature the general shape of the same bone in the adult. Look at Fig. 10. The middle part (A) is called

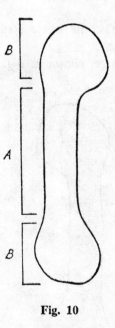

Fig. 10

the **diaphysis** and is often referred to as the **shaft.** The ends of the cartilaginous "bone" are enlarged (B) to form the **epiphyses** (singular:

epiphysis). The first appearance of actual bone is in the diaphysis shown at A in Fig. 11. It is therefore called the **primary ossification center**

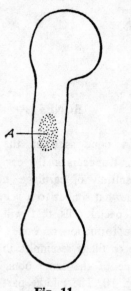

Fig. 11

—that is, the first site where cartilage is transformed into bone.

At a later stage, shown in Fig. 12, more of

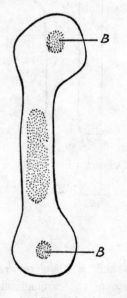

Fig. 12

the cartilage in the diaphysis had been transformed to bone, but now bone also appears in the enlarged ends, the cartilaginous epiphyses (B). The bone has not just expanded from the diaphysis into the epiphysis; the epiphysis is rather an independent center of bone formation. For this reason the small mass of bone in the cartilaginous epiphysis is called the **secondary ossification center.** At a stage (Fig. 13), which

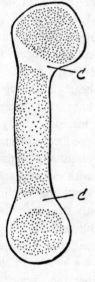

Fig. 13

resembles that found at birth, the diaphysis is entirely made of bone, and the secondary ossification center is expanding, but a disc of cartilage (C) separates the bone of the diaphysis from the bone of the epiphysis. This disc of cartilage is called the **epiphyseal disc** or **growth disc.** Increase in the *length* of the bone takes place *only* in the area of the growth disc. At a later period, sometimes during adolescence, sometimes in early adulthood, depending on the particular bone (Fig. 14), the cartilaginous epiphyseal disc is replaced by bone, and the bony diaphysis (A) is fused with the bony epiphysis (B). The lines of fusion (C) can often be seen in an X-ray picture of the bone. At this stage, increase in the length of the bone stops.

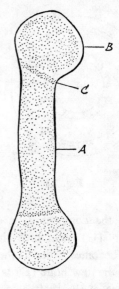

Fig. 14

a bony cylinder. The walls (B) are made of hard, dense bone called **compact bone.** The inside of the cylinder is filled with marrow. The bony epiphyses (C) are not hollowed out but are filled with strands of bone that crisscross and run in all directions. The small spaces between the bone strands are filled with marrow. The bone in the epiphyses is not dense and hard but relatively light in weight. This type of bone is called **cancellous bone.**

Marrow. Beginning during embryonic life, long before fusion has occurred, another process is taking place in the bones. In Fig. 15 the center

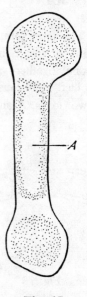

Fig. 15

of the bony diaphysis is slowly being hollowed out by absorption of bone (A). This excavation will form the **medullary** or **marrow cavity.** In the final adult bone (Fig. 16) the marrow cavity (A) extends throughout the shaft. It resembles

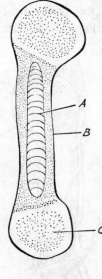

Fig. 16

Bone marrow is specialized tissue found in the marrow cavity of **long bones** as well as in cancellous bones. Marrow is of two types: (1) yellow, chiefly fat and therefore relatively inactive; and (2) red, which produces red blood cells as well as white blood cells of the granular series, for example neutrophils. At birth all marrow is red. However, with increasing age much of this red marrow is replaced by yellow marrow. In the adult limb bones only the upper ends of the femur and humerus contain red marrow. The bones of the head, vertebrae, and pelvis, contain red marrow throughout life as do the ribs and breast bone. In certain diseases in which there is excessive red blood cell destruction there is often some replacement of the yellow marrow with red. Red marrow may be partially destroyed by certain drugs taken over a long period of time. It would be useful, therefore, to know the normal distribution of red marrow in the various age groups. Some progress in gaining such knowledge is being made by the use of radioactive iron. Since iron is an essential element in the hemoglobin molecule, the administration of radioactive iron may result in increased radioactivity in those regions of the bone marrow where there is active red cell formation.

The skin of the bone, the **periosteum,** is shown at A in Fig. 17. It adheres closely to

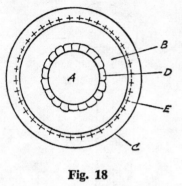

Fig. 18

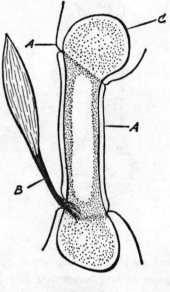

Fig. 17

the bone, especially at the points where muscles are attached to bone. The tendon of the muscle (B) is attached to the periosteum, and at this point the fibers of the periosteum penetrate the bone like nails, anchoring the muscle firmly. The rounded ends of the bone (C) are not covered with periosteum where it forms a joint or union with another bone.

Increase in the size of the bone takes place in two ways. The diameter or girth of the shaft increases by the process shown in Fig. 18. In this transverse section the marrow cavity (A) is

surrounded by the compact bone cylinder (B). The periosteum (C) surrounds the shaft. The marrow cavity becomes wider as the bone (D) is absorbed while the new bone (E) is laid down on the outer surface of the shaft by the periosteum. The increase in length of the shaft takes place, as shown previously, at the growth disc.

JOINTS

Fibrous Joints. The term **joint** means a union, or joining together, of two bones. The material that holds the bone together varies. The bones of the skull are joined by **fibrous tissue,** as shown in Fig. 19. The two bones, A and B, are covered

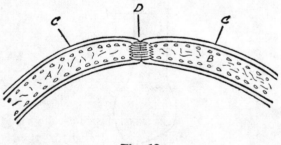

Fig. 19

by periosteum (C), and at the site of the union (D) the gap is filled with fibrous connective tissue. This is a **fibrous joint;** little movement, if any, occurs at such a joint.

Fibrous connective tissue is found throughout the body. It consists of threads of protein called collagen. In some areas the threads, or **fibrils,** are packed closely together, forming dense, fibrous tissue. Fibrous connective tissue is the material that holds the living cells of the body together, forming a fibrous skeleton, or framework. Sometimes it contains elastic fibers that can stretch like rubber bands. In other areas many fat-containing cells are found within the fibrous meshwork. A sheet of dense fibrous connective tissue is called a **fascia,** a term that will be used frequently.

In other joints the bones are bound together by cartilage instead of by fibrous connective tissue; these are called **cartilaginous joints.**

Synovial Joints. The most common type of joint found in the upper and lower limbs is the synovial joint. A diagram of such a joint is shown in Fig. 20. The two bones, A and B, are united by a synovial joint. The term **articulation,** meaning jointed, is frequently used to describe this union. The space between the two bones (C) is called the **cavity** of the joint. The space is much smaller than shown here. The binding material, in the shape of a sleeve, is called the **fibrous capsule** of the joint (D) and is continuous with the periosteum (E) of both bones. The ends of these bones in the living person are in contact with each other; therefore, to prevent wear and tear from constant friction, they are capped by a thin layer of cartilage (G). This cartilage is called **articular cartilage.** All surfaces within the joint cavity except the articular cartilage are covered with **synovial membrane** (F).

Synovial membrane is a very thin, smooth, shiny sheet of connective tissue and is responsible for the production of a viscous lubricating fluid for the joint cavity. The **synovial fluid** not only

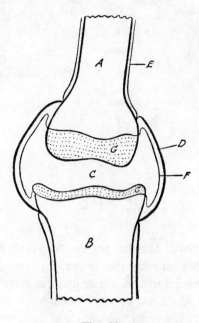

Fig. 20

reduces friction within the joint but also provides nutrition for the articular cartilage, which it bathes. The amount of synovial fluid produced is dependent on the physical activity of the joint. The stiffness and grating of joints observed after periods of inactivity are partly the result of a deficiency of synovial fluid.

Articular cartilage seems to thrive on intermittent pressure because it is thicker in lower limb joints, where there is greater weight on it than there is in the upper limb joints. It becomes extremely thin in people confined to bed for long periods of time. With increasing age there is a gradual decline in the healthy appearance of articular cartilage. Small cracks or fissures and areas of softening appear, and even small areas of bone may be completely bare of cartilage.

CHAPTER 3

MUSCLES

Introduction. Muscles are the "engines" of the body; they provide the power for movement. Those muscles that we use to perform movements of many kinds are called **voluntary** muscles because they are under the control of our own will. They make up more than 40 percent of the total body weight. The muscles of the heart and intestines are called **involuntary** because we cannot willfully control their contraction of relaxation. These muscles usually form the walls of hollow tubes; when they contract they push out the contents of the tube.

A voluntary muscle consists of a fleshy part or **belly** and ends in a cordlike sinew or **tendon.** The fleshy belly is attached to one bone while the tendon passes over a joint to become firmly attached to the adjoining bone. Shortening of the fleshy part of the muscle produces movement at the joint by pulling on the tendon. The tendon itself does not change in length.

There are several advantages to a tendon forming part of the muscle. If muscles were fleshy throughout their length, the joints of the wrist and ankle, for example, would have to be very thick and bulky. The tendons are also very strong and resistant to the rubbing or friction between the tendon and the moving bones of the joint.

Feel the fleshy belly of the "biceps" of your arm as you bend your elbow. Now try to lift the edge of a table with your palm and feel the "biceps." It should be firmer, for it is contracting more strongly. Follow the fleshy belly toward the elbow and feel its cordlike tendon. When you feel a muscle always make it perform work so that its strongly contracted belly may be more easily detected.

Types of Muscle. Three different types of muscle are found in the body. The most common type is **voluntary striated** muscle, which we use for any movements we want to perform. The name "striated" comes from the characteristic cross striations of the muscle fibers, as shown

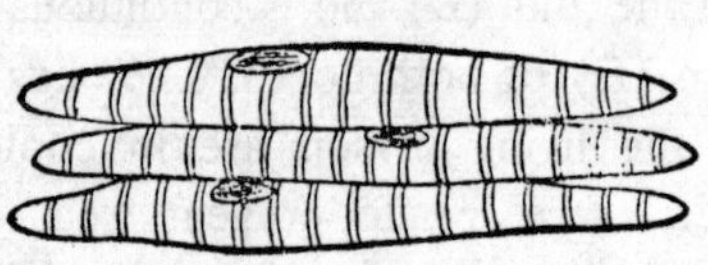

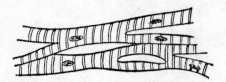

Fig. 21

in Fig. 21. These striations are caused by the arrangement of the protein molecules within the fibers that produce shortening or contraction of the fiber.

A type of striated muscle that is nonvoluntary is shown in Fig. 22. This is **cardiac** or **heart**

Fig. 22

muscle. It differs from voluntary striated muscle in that the fibers are jointed together, forming a network. (Another general term for network is **synctium.**)

The third type of muscle, nonstriated or involuntary, is shown in Fig. 23. This type of

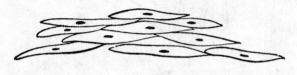

Fig. 23

muscle, often called **smooth** muscle, is found in the walls of blood vessels and in the walls of the alimentary canal. It is usually found as a sheath of muscle fibers surrounding a tubelike structure.

Voluntary striated muscle fibers are arranged in bundles within the muscle. A transverse section through a striated muscle (Fig. 24) shows a

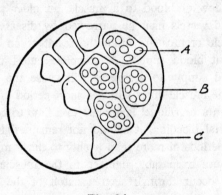

Fig. 24

group of muscle fibers (A) surrounded by fibrous connective tissue (B) to form a bundle. The many bundles surrounded by the fibrous connective tissue fascia (C) form the fleshy belly of the muscle.

The relation of the muscle bundles to the tendons is shown in Fig. 25. The muscle bundles (A) are surrounded and held together by the fibrous connective tissue (B), which is continuous with the fibrous connective tissue of the

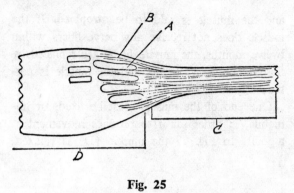

Fig. 25

tendonous part (C) of the muscle. The belly, or fleshy part, of the muscle is indicated by D.

Destruction of Muscles. Destroyed muscle fibers are never replaced. They do not have the ability to divide, and an individual has as many muscle fibers at birth as he will during his lifetime. The increase in the size of muscles of an adult compared with that of an infant is due to the increase in the *size* of the muscle fibers, not to an increase in their number. The increase in the size of muscle is called **hypertrophy.** Exercise causes hypertrophy of muscles.

NERVES, TENDONS, AND BLOOD SUPPLY

In Fig. 26 the nerve fibers separate within a

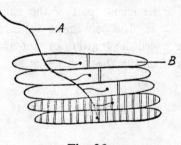

Fig. 26

muscle with a terminal branch of the nerve (A) going to each muscle fiber (B). If the nerve supply to a muscle is cut, a gradual change takes place in the muscle fiber. The fibers become smaller, the total size of the muscle decreases, and the muscle is said to be **atrophied.** If the muscle does not receive any nerve fibers within twelve months, the muscle fibers are replaced by fibrous connective tissue, and the muscle is permanently lost.

One end of the muscle must be fixed, or immobile, in order for it to produce movement at a joint. In Fig. 27 the muscle (D) is relaxed.

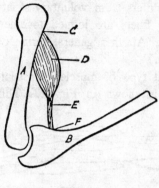

Fig. 28

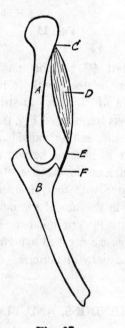

Fig. 27

It is attached to bone A at C and to bone B at F. The tendonous part of the muscle is indicated at E. The movement the muscle performs is called the **function** of that muscle. The function of muscle D is to bend, or flex, the bone

B. In Fig. 28 the muscle D has contracted, or shortened, by one-half of its original length to flex bone B. The length of the tendon remains the same. The attachment of the muscle at C, the non-moving or fixed end, is called the **origin.** The attached end, F, which moves during contraction, is called the **insertion** of the muscle.

Muscles have an extensive blood supply, and exercise is a powerful stimulus for increasing the flow of blood to a muscle. In older people blood vessels may be narrowed by disease, and muscle cramps are often the result of an insufficient blood supply. A complete cutoff of the blood supply to a muscle will cause the death of the muscle fibers within a short period of time. When a tourniquet is applied to a limb to control arterial bleeding, it should not remain tightened for periods of more than twenty to thirty minutes because irreparable damage to the muscle fiber may occur through such a cutoff of the blood supply.

CHAPTER 4

THE NERVOUS SYSTEM

Introduction. The human system consists of a brain and spinal cord which are the central receiving, integrating, and sending components of a communications network that spreads to all parts of the body. Some of the nerve fibers making up this network carry "information" from the skin, such as the feel of an object, pain from a cut, or the temperature of water. Other fibers transmit the degree of stretch of a muscle, while others relate the condition of the various organs of the body. This information is continually fed into the spinal cord and the brain. It may be stored in the brain or it may be used in determining the type of action that should be taken. If action is contemplated as a result of this information, the brain determines the intensity of the action, which will be governed by past experience, that is, the stored memory of a similar event. When the type of action has been decided upon, signals are sent out by way of other nerve fibers, which will stimulate the appropriate tissues to produce the desired action.

Nerve fibers relaying information to the brain and spinal cord are called **sensory nerves.** Those that transmit the commands of the brain to the tissues that perform the action are called **motor nerves.**

While the average brain weight in adults is almost three pounds, it reaches its adult size early in life. By seven years of age, the brain will have reached about 90 percent of its final weight. This precocious growth of the brain distinguishes it from most of the other organs. Although the average weight of a woman's brain is about 10 percent less than a man's, it should be obvious that brain weight by itself is not a useful index of an individual's mentality.

Neuron. The structural unit of the nervous system is the **neuron** (Fig. 29), which consists of a **cell body** (A) containing a **nucleus** (B) surrounded by **cytoplasm,** or **cell sap.** The typical neuron has two types of processes extending from the cell body. The first type of process is the **dendrite** (C), which is usually short and has a multiple branching pattern resembling the roots of a tree. The dendrite appears to be merely an extension of the cytoplasm of the cell body and plays a part in the nutrition of the cell. A nerve cell usually has many dendrites.

The other process is the **axon** (D). It may vary greatly in length, from a few millimeters to several feet, and from 1 to 20 microns in diameter. In contrast to dendrites, axons usually do not branch until they reach their terminations (E). Impulses are brought to the cell body by dendrites and carried from the cell body by axons. The speed of the impulse along an axon is related to its thickness. Impulses travel faster along axons of larger diameters. While most nerve cells have only one axon, sensory neurons have two axons (G) arising via a common stem from the cell body (F). This type of neuron carries sensations, for example, from the skin (H) to the spinal cord and brain (I).

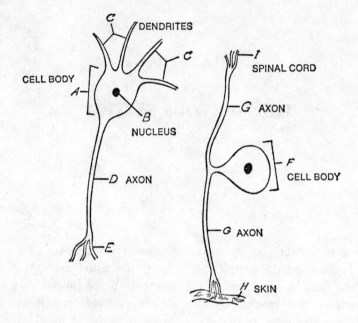

Fig. 29

Neurons in the brain and spinal cord are described as being within the **central nervous system.** All other neurons are within the **peripheral nervous system.** A typical axon in a peripheral nerve (Fig. 30) consists of the **axis cylinder** (A) enclosed by a fatty insulation called **myelin** (B). The myelin covering is interrupted at intervals along the axis cylinder, forming the **nodes of Ranvier** (C). These nodes are believed to govern the speed of the impulse along the axon. Axons in peripheral nerves are additionally surrounded by a sheath of cells called the **Schwann cells** (D) external to the myelin layer. These sheath cells are not found as such around axons in the central nervous system.

Impulses. If a cell body is stimulated, an impulse having many properties characteristic of an electric current passes along the axon to its terminal branches, which end as small sacs called **vesicles** (E). At this point the axon is in contact but not in continuity with either the dendrite of another neuron or a muscle fiber. The impulse traveling along the axon (F) causes the release of a chemical substance contained within

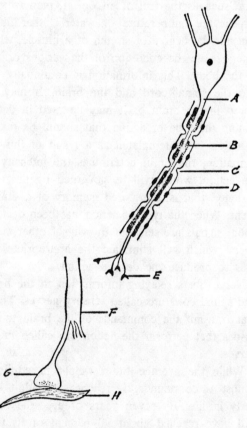

Fig. 30

the vesicle (G), which causes the muscle fiber (H) to contract. This point of contact between the terminal ends of the axon and the muscle fiber or dendrite is called a **synapse.**

Destruction of Nerves. Nerve cell bodies are incapable of dividing; therefore, when a nerve cell is injured or destroyed, the loss is permanent. Destruction of axons in the central nervous system is also permanent. This is in contrast to the sequence of events that occurs when an axon in a peripheral nerve is severed (Fig. 31A). The

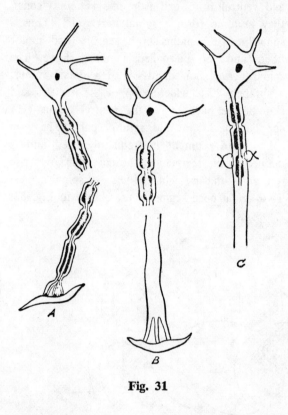

Fig. 31

axon distal to the cut degenerates and the axis cylinder is removed along with the myelin by phagocytic cells, leaving an empty tube consisting of sheath cells (Fig. 31B). The cut end of the axis cylinder connected with the cell body begins to grow down the hollow sheath and will finally reach the end organ originally supplied by the axon (Fig. 31C). Thus the cut ends of the nerve must be carefully sutured or joined together, so that no gap is left between the cut

ends and so that the hollow sheath is in the correct position to receive the growing nerve fiber.

CRANIAL AND SPINAL NERVES

There are twelve pairs of **cranial nerves** and thirty-one pairs of **spinal nerves.** The cranial nerves arise from the brain. Of the spinal nerves, eight pairs arise from the cervical part of the spinal cord, twelve from the thoracic, five from the lumbar, five from the sacral, and one from the coccygeal segments of the cord. The spinal nerves are named according to their relationship to a particular vertebral body. The first cervical nerve appears between the base of the skull and the atlas, or first, cervical vertebra. The eighth cervical nerve emerges between the seventh cervical vertebra and the first thoracic vertebra. Thereafter, a spinal nerve is named from the vertebra that lies immediately above it as it exits from the vertebral canal. For example, the eleventh thoracic spinal nerve emerges between the eleventh and twelfth thoracic vertebrae.

Sensory and Motor Tracts. The spinal cord (Fig. 32) consists of a small central canal (A)

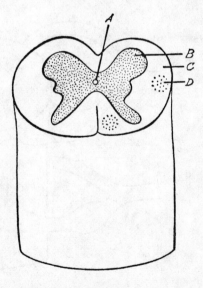

Fig. 32

surrounded by masses of nerve cell bodies that, when seen in a section of a fresh cord, have a gray color. Hence these cellular masses are called **gray matter** (B). External to the gray matter is an area of **white matter,** which consists of axons (C). These axons are grouped into **tracts** (D), which ascend and descend in the cord. Ascending fibers going to the brain are called **sensory tracts.** Other axons descending from the brain to end in the gray matter of the spinal cord are called **motor tracts.** Sensory fibers from the periphery of the body entering one side of the spinal cord will cross to the opposite side of the cord before they end in the brain. In a similar manner, motor fibers arising from one side of the brain soon cross to descend in the opposite side of the spinal cord. Thus, destruction of the motor areas in the left side of the brain results in the loss of voluntary movement in the right side of the body.

Spinal Nerves. The anterior projection of gray matter in the spinal cord (Fig. 33) is called the **anterior horn** (A). Both anterior horns contain cell bodies from which axons arise to exit from the cord and innervate voluntary muscle. The **posterior horns** (B) contain cell bodies upon which sensory axons of peripheral nerves synapse. Motor axons leaving the cord are grouped together to form the **ventral root** (C), while the group of sensory axons entering the cord form the posterior, or dorsal, root (D). The visible swelling, or **ganglion** (E), on the dorsal root contains the cell bodies of those sensory axons entering the spinal cord. Just before the dorsal and ventral roots exit from the vertebral canal, they unite to form a **spinal nerve** (F). Thus a spinal nerve contains both sensory and motor fibers and can be called a mixed nerve. The spinal nerve soon divides into a **dorsal ramus,** or branch (G), which supplies the muscles and skin of the back, while the **ventral ramus** (H) supplies the lateral and anterior parts. The term **spinal cord segment** is frequently used and is defined as the segment of the cord that gives rise to a pair—that is, right and left—of spinal nerves. Two spinal cord segments are shown in Fig. 33.

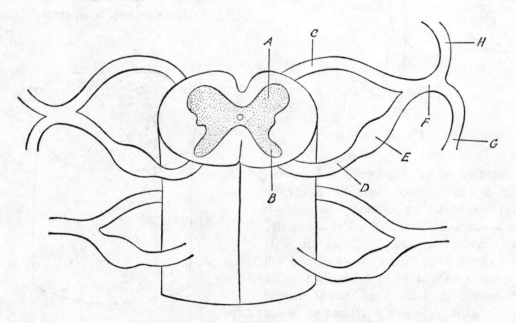

Fig. 33

MOTOR REFLEXES

Knee Reflex. It will be possible now to visualize the nerve pathways involved in a simple reflex (Fig. 34), a basic functional property of the nervous system. From a sitting position, cross your legs at the knees and sharply tap the tendon just below the knee cap (A). As the knee is already bent and the tendon somewhat stretched,

the force of the tap further stretches the tendon, resulting in a sudden kick or extension of the leg that is impossible to control. Specialized nerve endings in the tendon (B) are stimulated when the tendon is stretched, and the impulse travels along the sensory axon (C) in the spinal nerve to enter the spinal cord through the dorsal root (D). The axon synapses with a cell body in the posterior horn (E). This cell body is a com-

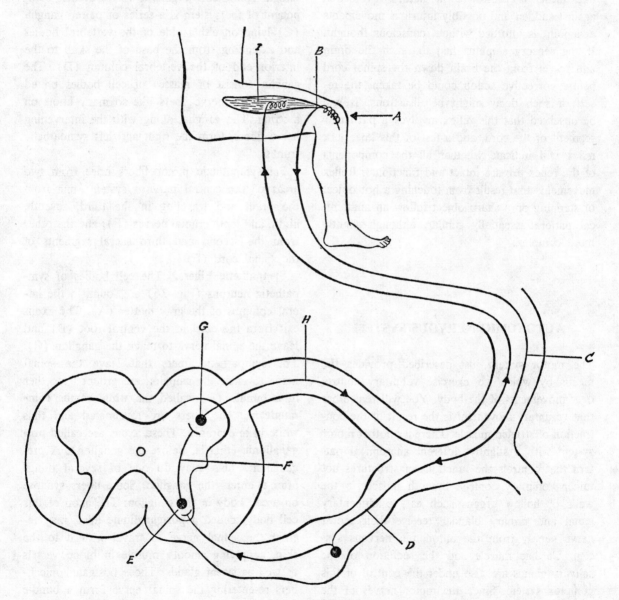

Fig. 34

ponent of the connecting neuron (F), which synapses in turn with a cell body in the anterior or ventral horn (G). Stimulation of this motor cell results in an impulse that travels along its axon (H) through the ventral root to the spinal nerve, where it reaches the muscle (I) and causes it to contract. Contraction of the muscle straightens the leg.

Protective Nature of Reflexes. This reflex, like most, is protective in nature. Protection against sudden and possibly injurious movements at a joint is afforded without conscious thought. If the sensory stimulus had to reach the brain and travel from the brain down the spinal cord before corrective action could be taken, the result of such delay might be disastrous. It will be observed that this reflex involves a particular segment of the cord, and a test of this knee jerk reflex will indicate whether all the components of the reflex arc are intact and functional. Reflex movements that result from touching a hot object or stepping on a sharp object follow an anatomical pattern essentially similar, although slightly more complex.

AUTONOMIC NERVOUS SYSTEM

The motor system just described provides the means by which we exercise voluntary control over movements of the body. You will remember that voluntary movement is the result of the contraction of striated muscle. There is another motor system with a slightly different anatomical pattern that controls the functions of structures not under voluntary control. Smooth muscle in the walls of hollow organs such as the alimentary canal and urinary bladder receive their motor nerve supply from the autonomic nervous system. Cardiac muscle and the secretion of the salivary glands are also under the control of this complex system. Since the motor nerves of the autonomic system follow one of two anatomical patterns, and stimulation of them produces two different types of physiological response, this system is subdivided into the **sympathetic** and **parasympathetic** systems.

Sympathetic System. A basic plan of both systems is outlined in Fig. 35, in which a diagram of the brain and spinal cord is shown. The sympathetic system consists of paired neurons (A) that arise from all the thoracic and the first two lumbar spinal cord segments (B). Another component of the system is a series of paired ganglia (C) lying on either side of the vertebral bodies and extending from the base of the skull to the inferior end of the vertebral column (D). The ganglia consist of masses of cell bodies united vertically by nerve fibers like so many knots on a string. The ganglia, along with the intervening nerve fibers, form the right and left sympathetic trunks.

Parasympathetic motor fibers arise from two areas of the central nervous system: one from the brain and traveling in the third, seventh, ninth, and tenth cranial nerves (E); and the other from the second and third sacral segments of the spinal cord (F).

Sympathetic Fibers. The cell bodies of sympathetic neurons (Fig. 36) are found in the lateral columns of the gray matter (A). The axons exit from the cord in the ventral root (B) and leave the spinal nerve to reach the ganglion (C). The sympathetic fibers that leave the spinal nerve to enter the ganglion are grouped together in a bundle (D) called the **white ramus communicans.** The fibers are myelinated and thus white in appearance. These axons are called **preganglionic** (that is, before the ganglion). A preganglionic fiber may do one of several things after it enters the ganglion. Some fibers synapse on a cell body in the ganglion. The axon of this cell body, called a **postganglionic** fiber, may reenter the spinal nerve to travel with it to the skin, supplying smooth muscle in blood vessels as well as sweat glands. Those postganglionic fibers re-entering the spinal nerve form a bundle called the **gray ramus communicans** (E). Post-

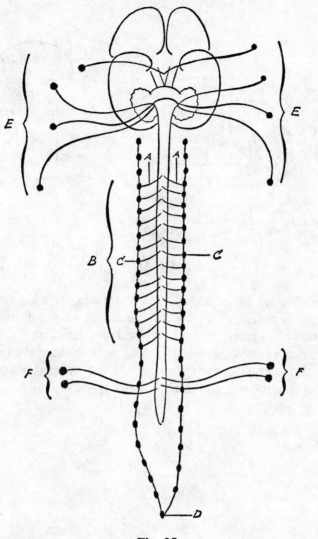

Fig. 35

ganglionic fibers are not myelinated and hence have a gray color. Other preganglionic fibers may pass through the ganglion to travel up or down the sympathetic trunk before synapsing. Still other preganglionic fibers may leave the ganglion as well as the sympathetic trunk to travel to special sympathetic ganglia associated with blood vessels in the abdominal cavity (F).

As a general rule, sympathetic ganglia are relatively far removed from the organ they supply—that is, the postganglionic sympathetic fibers are relatively long. In contrast, parasym-

pathetic ganglia are relatively close to the organ supplied—that is, they are relatively short.

The sympathetic nervous system prepares an individual for "fight or flight." Stimulation of this system results in dilation of the pupils, an increase in the rate and strength of contraction of the heart, a decrease in peristalsis of the gastrointestinal tract, and a diminished blood flow in the skin and gastrointestinal tract in order to increase the blood flow to voluntary muscles. At the same time, the nerves of this system cause the adrenal gland to secrete adrenalin into the

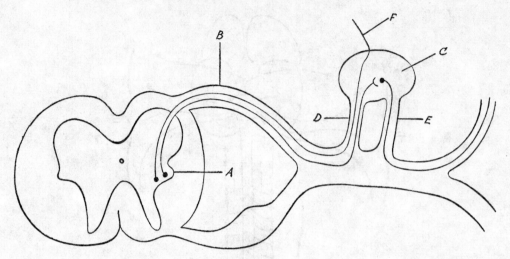

Fig. 36

blood stream, thus reinforcing the physiologic functions of the sympathetic fibers.

Parasympathetic System. The parasympathetic system produces somewhat opposite effects. It produces constriction of the pupil, adjusts the lens of the eye so that near objects are more clearly seen, increases the flow of saliva, decreases the rate and strength of contraction of the heart, increases the flow of digestive juices and peristaltic movements of the gastrointestinal tract, and permits emptying of the bowel and bladder.

CHAPTER 5

BLOOD VESSELS AND LYMPHATIC SYSTEM

Introduction. The blood vessels form a closed system of variously sized tubes which extend to all parts of the body. The tubes carrying blood away from the heart are called **arteries,** while those that return blood to the heart are **veins.** The force or pressure required to keep the blood circulating is provided by a muscular pump— the heart. The blood functions to carry oxygen and nutritive materials to the tissues and to remove the waste products of the tissues. Many hormones also reach the tissues by way of the blood stream. Protective substances in the blood, called **antibodies,** are carried to infected areas by the blood vessels.

The blood within the vessels consists of cells and a relatively clear liquid called **plasma.** The cells are very small packets submerged in the plasma. Most of the cells are red, for they contain a reddish-colored substance called **hemoglobin,** which gives blood its characteristic color. The remaining cells, called white blood cells, or **leukocytes,** are concerned with protecting the body against infection.

The red blood cells carry oxygen to the tissues and, after discharging their cargo, pick up carbon dioxide, a waste product of the tissues. The red blood cells now loaded with carbon dioxide are carried by the vessels to the lungs, where the carbon dioxide leaves the red cells to enter the air sacs to be finally exhaled into the atmosphere. After discharging the carbon dioxide the red blood cells within the lung vessels take

on a new cargo of oxygen and the cycle is repeated.

Those vessels carrying blood to and from the lungs form the **pulmonary circulation.** Thus the heart is really two pumps situated side by side. One pump circulates the deoxygenated blood to the lungs to allow the red cells to discharge carbon dioxide and take on oxygen. The other pump sends the oxygenated blood to all the other tissues of the body, where the red cells give up their vital oxygen to the tissues and take on the unwanted carbon dioxide. The heart pumps out almost five quarts of blood every minute in a normal resting person, but it can increase its output about five times that amount during strenuous exercise.

Kinds of Vessels. An appreciation of the difference between **arteries** and **veins** may be gained by feeling the **radial pulse** about an inch proximal to the base of the thumb as well as the veins on the dorsum of your hand. An artery pulsates with every contraction of the ventricle of the heart; its walls suddenly expand and then recoil. No pulsation can be felt in a vein. Press on the artery and the vein to shut off the blood flow. Much greater force is required on the artery than on the vein. The chief reason for this difference is that the pressure of the blood flow in the artery is much higher than that in the vein. Because of the higher pressure in the artery, the arterial wall is thicker than the wall of the vein. The degree of thickness of a vessel wall is dependent on the relative amounts of smooth muscle surrounding the **lumen,** the open part of the tube. These features characterize arteries and veins in all parts of the body.

As an artery, for example, the **aorta,** is followed from the heart, not only does the caliber of the lumen become smaller, but the walls also gradually become thinner. The arterial vessels that connect with the **capillaries** are the **arterioles,** an important but puzzling component of the arterial system. The capillaries connect in turn with little veins, or **venules,** that become progressively larger and finally empty into the heart. Large arteries are yellowish in color because of the elastic tissue in their walls. It is this elastic tissue that is responsible for the expansile property of an artery. The pulse wave in an artery, although initiated by the contraction of the ventricle, proceeds along the artery at a faster rate than the flow of blood. It aids the heart by smoothing out the flow of blood, since the elastic recoil will also push blood along the artery. The expansion of the arterial wall also tends to buffer the pressure in the artery; in degenerative disease of arteries, such as **arteriosclerosis,** or hardening of the arteries, there is a much greater rise and fall of the pressure with each contraction and relaxation of the ventricle.

Capillaries. Capillaries have the thinnest walls in the vascular system. They are only a layer of flattened cells called **endothelium,** which forms a tube with a diameter that will barely allow the passage of a red blood cell. These walls allow fluids containing protein to pass from the capillaries to the spaces between cells, the **intercellular space,** as well as from the intercellular space back to the capillaries. Because of the slightly higher pressure, the flow at the arteriolar end of the capillary is from the capillary, carrying nutritive substances in addition to oxygen to the cells. Cellular waste products and carbon dioxide enter at the venule end of the capillaries.

Arteriole. In addition to the endothelial lining, the arteriole has an outer circular coat of smooth muscle. The muscle is innervated by the sympathetic nervous system, which controls the caliber of the lumen and thus the flow of blood to the capillaries.

Capillary Bypass System. In many areas of the body arterioles connect directly with venules and so bypass the capillary bed. There are several advantages in having a capillary bypass system. If all the capillaries in the voluntary muscles of the body were filled with blood, they would contain almost all the available blood in the body. When you consider the capacity of all the capillaries in the body, you can appreciate the magnitude of the surface area of the capillary bed. It is obvious that many capillaries must be shut off for varying periods of time. The blood flow through the capillary bed is controlled by the constriction of arterioles as well as by the arteriolar-venule bypass. The major part of the body heat is lost by convection from the capillary bed in the skin. When your hand is exposed to cold, the skin becomes pale, indicating empty capillaries. Capillary bypass is a mechanism by which the body can conserve heat. The reverse takes place when the rate of heat loss from the body must be increased. During digestion, the flow of blood through the capillary bed in the walls of the alimentary tract is greatly increased, but during quiescent periods the capillary bed is bypassed. A similar pattern is followed in many other organs. Generalized constriction of arterioles increases resistance to blood flow in the larger arteries, producing an increase in blood pressure, or **hypertension.** Surgical removal of sympathetic nerves to arterioles will decrease peripheral resistance and thus lower blood pressure.

COLLATERAL CIRCULATION

In most parts of the body more than one artery supplies the same area. Not only does the capillary bed of one artery connect with that of an adjacent artery, but larger branches of one artery will also directly connect with a neighboring artery. This end-to-end joining is called an **anastomosis** and will provide an alternate route for the blood to the area if one artery is closed off. This alternate routing for blood is called **collateral circulation.**

Arterial Blocks. Many organs have an extensive collateral circulation, but there are several factors that will determine whether collateral circulation is sufficient to maintain the organ if the main artery is blocked. If the vessel block is sudden, the collateral vessels may not have sufficient time to enlarge before the tissue dies. Gradual narrowing of an artery is a stimulus for the enlarging of the collateral vessels. Blockage of old arteries presents a more dangerous problem than that of young arteries. The site of the arterial block is of importance. If the block is between the heart and the point where the anastomosis takes place, the collateral vessels will not receive blood either. If an artery is surgically tied off, the site chosen is usually just beyond the area where the anastomosis between two adjacent arteries occurs.

Most organs have considerable collateral circulation, but several do not. The brain and kidney arteries do not anastomose, so they are called **end arteries.** Blockage of one artery is therefore followed by death, or **necrosis,** of the area supplied. This necrotic area is called an **infarct.** An infarct in the brain tissue is much more serious than one in the kidney. Because of the specialized function of each area in the brain, no other area can take over for the necrotic part. Since we have more kidney tissue than is needed for survival, the death of a small area is not fatal to the organism. Even a part of one kidney may be adequate for survival, which implies that the kidney has a large functional reserve.

While the small arteries in the heart muscle show some anastomosis, it is usually not sufficient to maintain the muscle if the main artery is blocked. They are regarded as functional end arteries. Whether or not the blockage of an artery to the heart muscle will be fatal is usually determined by the size of the vessel involved and hence the size of the necrotic area, that is, the infarct.

Valves. Veins show much more extensive collateral circulation than do arteries, and functional difficulties following blockage are rare. Many veins have projections of their endothelial lining, called **cusps,** which project into the lumen of the vessel. Two or more cusps are arranged to form a **valve** that permits the blood to flow in only one direction—toward the heart. Not all veins have valves. Veins in the abdominal cavity and most veins in the head and neck lack valves because the pumping action of the abdominal muscles and diaphragm, the negative pressure in the thoracic cavity, and gravity preclude the need for valves. In the lower extremities the flow of blood in the veins is from the superficial veins (those just deep to the skin) to the deep veins (those deep to the muscles). Small anastomotic channels protected by valves insure that the blood travels from the superficial to the deep veins. If for some reason the edges of the cusps do not meet, the valve is inefficient and blood can then flow in the reverse direction (from the deep to the superficial veins). The extra load thrown on the superficial veins combined with the effect of gravity causes the superficial veins to dilate. As a result, more valves become incompetent, and a longer column of blood must be supported by the vein wall. The end result is a dilated, twisted, or tortuous vein, called a **varicose vein,** that cannot efficiently remove the blood from the skin and superficial structures. Since the nutrition of the skin is adversely affected by the inefficient blood circulation, a varicose vein interferes with the normal protective functions of the skin. Methods used to treat varicose veins include the surgical closure of the connections between the deep and superficial veins, the removal or obliteration of the dilated vein, or compression of the vein by an elastic support. Standing for long periods of time allows gravity to magnify the condition, while elevation of the extremity above the level of the heart is beneficial.

William Harvey was the first to describe the circulation of the blood correctly. He showed the function of the valves in veins by a simple but classic experiment. Allow your arm to hang downward until the veins on the back of your hand become distended. Choose a vein that joins another vein proximally. With two fingertips, compress the vein over its middle part. Move one fingertip proximally along the vein to its junction with the other vein so as to press the blood out. Now remove the distal finger and you will see that the empty vein rapidly fills up to the proximal fingertip whether or not a valve is present. Repeat the maneuver but lift the proximal finger. If a valve is present at the venous junction, the empty segment of the vein remains collapsed; if there is no valve, the empty segment will fill very slowly from above.

LYMPHATIC SYSTEM

Perhaps the most fascinating story yet to be told in biology is the function of the lymphatic system. Although the anatomical plan of this system was presented many years ago, its function has not yet been understood. The **lymph vessels** start as blind channels in the intercellular spaces. Fluids, protein particles, and inert particles such as carbon are readily taken into the lymph vessels. Small vessels join adjacent ones to form larger channels on their ultimate destination to the root of the neck, where they empty into large veins. Interrupting the channels are **lymph nodes,** which in some instances seem to act as filters for materials carried in the lymph vessels. Lymph nodes vary greatly in size. Normal lymph nodes may be felt in the groin. Swollen lymph nodes may be palpated in the neck of a person with infected tonsils or in the **axilla** (armpit) of a person with an infected finger. The trapping of bacteria or bacterial products by the lymph node filters appears to be a natural defense mechanism.

In contrast, lymph vessels act as natural carriers for cancer cells. Such cells separating from a cancerous growth in the breast will travel along lymphatic vessels and when caught in a lymph node in the axilla will multiply there. The node will enlarge as a result, and its normal function will be destroyed. Succeeding cancer cells will bypass the damaged node to involve other nodes farther along the lymphatic chain. For this reason, lymph nodes receiving lymph vessels draining a cancerous area should be removed along with the cancer. Sometimes it is technically impossible to remove surgically all of the involved nodes, so the cancer frequently reappears in these nodes. It should be obvious that the early detection of cancer, followed by the surgical removal of the growth along with the accessible lymph nodes, would give the best long-term results.

Lymphocytes. Perhaps the most important function of lymph nodes is the production of **lymphocytes,** a type of white blood cell. For many years the lymphocyte was a cell with an unknown function. Recent evidence strongly suggests that the lymphocyte is involved in **antibody** formation, although its exact sequence in the chain of events is unknown. An antibody, a complex protein molecule, is formed in the body and aids in the destruction of all foreign material gaining entrance to the body, whether bacteria, virus, pollen, or organs transplanted from another person.

Organ transplantation is now technically possible, but destruction of the organ by antibodies frequently follows. Organs exchanged between identical twins are not destroyed by antibodies because they have developed from identical genetic material and hence are not foreign to either twin. If the lymphoid tissue is destroyed or paralyzed by X rays or chemicals, grafted organs survive for much longer periods, but the host is unable to protect himself against bacteria or viruses. When the complete story of lymphocyte function is unfolded, then control of antibody production may be within reach. Perhaps the lymphocyte will supply the vital link in this process.

CHAPTER 6

HEAD AND NECK

SKULL

Bones. The individual bones of the skull can be more easily distinguished in an infant's skull (Fig. 37) than in an adult's. In this lateral

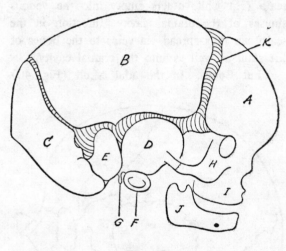

Fig. 37

view the **frontal** (A), **parietal** (B), and **occipital** (C) bones are shown. The **temporal** bone consists of four parts: the thin, flat **squamous** (D), the **mastoid** (E), the **tympanic** (F), and a **petrous** portion of which only the pointed **styloid process** (G) can be seen from this aspect. The **zygomatic** bone (H), forming the prominence of the cheek, joins the squamous part of the temporal bone to the **maxillary** bone (I). The **mandible,** or lower jaw, is shown at J.

Membrane Bones. Many of the bones of the roof and sides of the skull are called **membrane bones** because they are formed within a tough, fibrous membrane. Membrane bone (Fig. 38)

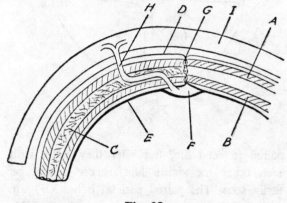

Fig. 38

consists of an outer layer (A) and an inner layer (B) of hard, compact bone, with the space between the two layers filled with cancellous bone (C). The cancellous bone in the skull contains red marrow throughout life.

Most of the bones forming the floor of the skull, however, are formed in cartilage, as are the bones of the limbs (see Chapter 2). The parietal bone, shown at B in Fig. 37, is an example of a membrane bone; the membrane not yet invaded by bone is shown at K. The bone increases in size by laying down new bone around its edge so that finally the membrane

(K) can no longer be seen. When growth is completed and adjacent bones have become interlocked, the line of union has a jigsawlike pattern called a **suture line.**

Sutures. Many of the skull bones are paired, as seen in Fig. 39. The two frontal bones are

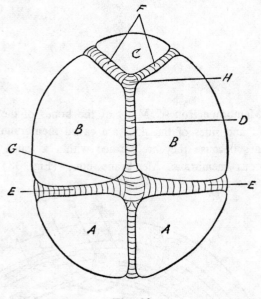

Fig. 39

paired in fetal life, but after they unite with each other the suture line usually cannot be easily seen. The paired parietal bones (B) are separated from each other by a membrane (D), and when they unite, the longitudinal interlocking line is called the **sagittal suture.** Similarly, the membrane (E) between the parietal and frontal bones will mark the site of the **coronal suture.** The membrane between the parietal bones (B) and the occipital bone (C) indicates the shape of the future **lambdoid suture,** a name taken from the shape of the Greek capital lambda, λ.

Fontanelles. In the fetus and infant two relatively large areas of membrane are present in the skull roof. One is at the junction of the two parietal and two frontal bones. This area, G in Fig. 39, is called the **anterior fontanelle,** meaning little fountain. It can easily be felt in the

infant, and pulsation corresponding to the heart beat can be seen as well as felt. By the time a child is two years old, the adjacent bones have closed the gap. Before closure of the anterior fontanelle, blood may be obtained by inserting a needle through the membrane into the large venous sinus lying directly inferior. The **posterior fontanelle** (H) closes at birth or shortly after.

Adhering closely to the outer layer in Fig. 38 is periosteum (D) continuous with a periosteal-like layer (E) through the suture line (G). The layer (E) covering the inner table is part of the **dura mater,** the outer membranous covering of the brain. In certain areas the dura mater splits to enclose a **venous sinus** (F). These venous sinuses drain most of the blood from the brain. The cancellous bone layer between the tables of compact bone contains a network of veins; some of them communicate with the veins of the scalp (H) while others empty into the venous sinuses of the cranial cavity. Infection in the scalp can thus spread via veins to the bones of the skull as well as into the cranial cavity.

Adult Bones. In the adult skull (Fig. 40)

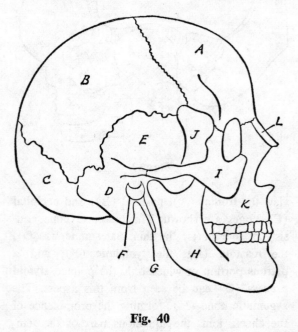

Fig. 40

the membrane between the individual bones has

been replaced by bone, and only the suture line can be seen. The coronal suture separates the frontal bone (A) from the parietal bone (B). One limb of the lambdoid suture separates the occipital bone (C) from the parietal. Four parts of the temporal bone can be seen: the squamous part (E), the mastoid with its process (D), the tympanic part (F), which forms most of the bony external auditory canal, and the styloid process (G) of the petrous temporal. The bone (J) that adjoins the temporal, parietal, and frontal bones is called the **greater wing of the sphenoid.** The right and left greater wings are attached to the body of the sphenoid bone, which lies in the central part of the floor of the skull.

Bones of the Face. The zygomatic bone (I) is attached to the temporal bone posteriorly, and anteriorly to the maxilla (K) and frontal bones. The short **nasal bone** (L) is paired. The mandible (H) pivots against the inferior surface of the temporal bone during opening and closing of the mouth.

The mandible, as seen from the lateral aspect (Fig. 41), consists of a **ramus** (A) and a **body**

forms part of the jaw joint, and a neck. The **coronoid process** (D) receives the insertion of the large **temporalis** muscle.

The angle of the mandible (E) is formed where the posterior border of the ramus meets the inferior border of the body. The bony opening in the body of the mandible (F) is an exit for the terminal branch of the inferior alveolar nerve, which runs within the bone to supply the lower teeth. The bone of the mandible is hard and dense, so a local anesthetic placed near the root of a tooth has little chance of reaching the branch of the inferior alveolar nerve supplying that tooth. Thus, the nerve itself must be anesthetized before it enters the mandible. The bone of the maxilla is much less dense and has many minute openings. An upper tooth is more easily anesthetized by injection of the anesthetic into the gum near the root of the tooth.

On the medial surface of the mandible (Fig. 42), the opening in the bone (A) through which

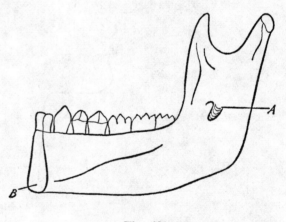

Fig. 42

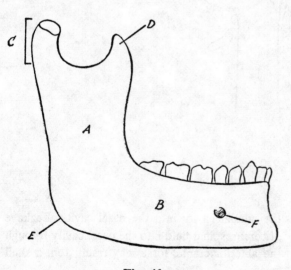

Fig. 41

(B). The upper end of the ramus has a **condyloid process** (C) consisting of a head, which

the inferior alveolar nerve enters is shown. The nerve is usually anesthetized in this area—that is, before it enters the bone. The right and left mandibles are joined by fibrous tissue at the midline of the chin (B), but in adults the fibrous tissue has been entirely replaced by bone.

CRANIAL CAVITY

Fossae. Since the skull is molded on the brain, the floor of the cranial cavity consists of three fossae, or depressions, each at a different level and each containing a different part of the brain (Fig. 43). The **anterior cranial fossa** (A) is formed by the bony roof of the orbit; the frontal lobe of the brain rests upon it. At a slightly lower level the **middle cranial fossa** (B) extends to the posterior surface of the skull. Part of the floor is bony, and part is formed by a tough membrane (D) called the **tentorium cerebelli.** The temporal lobe of the brain fits into the anterior end of the middle fossa, while the occipital lobe of the brain rests on the tentorium. The **posterior cranial fossa** (C) is that part of the cranial cavity lying inferior to the tentorium. It contains the cerebellum and the other parts of the hindbrain.

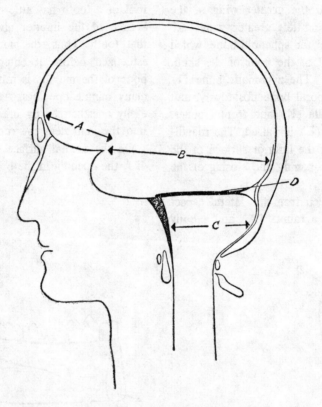

Fig. 43

When the roof of the skull, or **skull cap,** is removed along with the brain, the bones forming the floor of the cranial cavity are seen (Fig. 44). Anteriorly the floor of the anterior cranial fossa is formed by the **orbital plate** of the frontal bone (A). In the midline the **cribriform plate** of the **ethmoid** (B) allows the olfactory nerves—the nerves serving the sense of smell—to pass from the cranial cavity into the nasal cavity. Leakage of cerebrospinal fluid into the nasal cavity through the anterior cranial fossa may result from a skull fracture.

Sphenoid. The sphenoid is viewed from above, showing the **lesser wings** (C) separated from the **greater wings** (F) by the superior orbital fissure. The **pituitary fossa** (D) on the body of the sphenoid is surrounded by four bony prongs. The two anterior prongs are part of the lesser wings and are called the **anterior clinoid proces-**

45. The bone is viewed from its posterior end. It consists of a body (A), which contains the **sphenoidal air sinus** (B) and two pairs of wings.

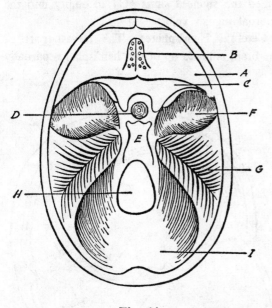

Fig. 44

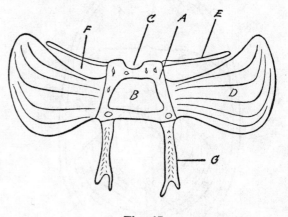

Fig. 45

ses. The posterior clinoid processes are part of the **dorsum sellae** (E), the raised posterior wall of the body of the sphenoid. The paired pyramid-shaped bony structures (G) are the petrous parts of the temporal bone. One side of the pyramid faces downward and medially and forms part of the side wall of the posterior cranial fossa. Another side faces upward and laterally and is part of the middle cranial fossa. The edge between these two sides gives attachment to the tentorium cerebelli. The remaining side of the petrous bone can be seen on the inferior aspect of the skull. Looking into the posterior cranial fossa, the large **foramen magnum** (H) surrounds the continuation of the brain stem, the spinal cord. The foramen magnum is surrounded by the occipital bone (I).

The sphenoid bone is extremely difficult to visualize from a diagram of articulated skull bones. If separated from the other bones, its parts would resemble the diagram shown in Fig.

A depression in the roof of the body (C) receives the **pituitary gland,** which hangs by a stalk from the brain. A pair of wings spreads out from each side of the body. The lesser wing (E) is separated from the greater wing (D) by a fissure, or gap, in the bone, the **superior orbital fissure** (F). Most of the vessels and nerves going to structures in the orbit pass through this fissure, although the optic nerve, which has its own canal, does not. From the inferior surface of the body two leglike processes (G), called the **pterygoid processes,** project downward.

THE BRAIN

Supporting Structures. The fibrous supporting structures of the brain, formed by the **dura mater,** are shown in Fig. 46. The two structures are the **falx cerebri** (C), which dips between the cerebral hemispheres, and the **tentorium cere-belli** (D), which separates the cerebrum from the cerebellum. To place the falx cerebri in its correct position, superimpose on the areas A and B the corresponding areas in the cranial cavity so that the falx is at right angles to the tentorium. The falx contains a large **superior sagittal venous sinus** (E) and a small **inferior**

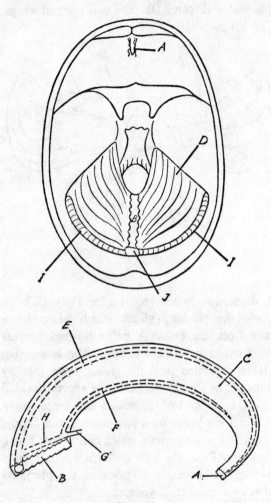

Fig. 46

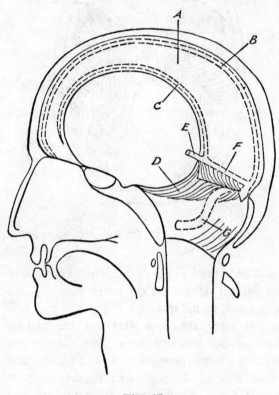

Fig. 47

that the straight sinus lies at the junction of the tentorium cerebelli (D) with the falx cerebri. The transverse sinus runs horizontally, then dips into the posterior cranial fossa, where it is called the **sigmoid sinus** (G) to empty into the internal jugular vein.

Cerebral Hemispheres. The largest part of the brain is made up of two hemispheres partially

sagittal sinus (F). A large vein from the brain (G) joins the inferior sagittal sinus to form the **straight sinus** (H). Blood from these sinuses now enters the transverse sinus, right and left (I), at J. Note the notch in the tentorium cerebelli that encircles the midbrain.

As seen in their normal relationships (Fig. 47), the superior (B) sagittal sinus of the falx cerebri (A) drains into the transverse sinus. The inferior sagittal sinus (C), in common with the great cerebral vein (E), however, reaches the transverse sinus via the straight sinus (F). Notice

separated by a deep midline groove. The right and left hemispheres constitute the **cerebrum.** The lateral surface of the left cerebral hemisphere is shown in Fig. 48. The surface of the cerebrum is thrown into folds called **gyri** (singular, gyrus), which are separated from each other by grooves called **sulci** (singular, sulcus). The convolutions of the cerebral hemispheres provide maximum surface area with a minimal increase in volume.

Cerebral Cortex. The cerebrum consists of

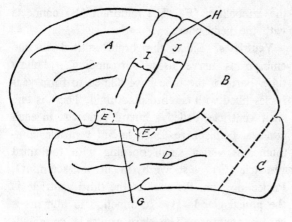

Fig. 48

an outer layer of gray matter composed of nerve cells, which are the receiving and sending stations for body functions. This outer layer of gray matter is called the **cerebral cortex.** Deep to the cortex is the larger mass of white matter, which consists of nerve fibers going to or leaving the cortex. Within the white matter are small masses of nerve cells, called nuclei, which send or receive fibers of cerebral nerves. The architecture of the cerebellum is somewhat similar. In the remaining parts of the brain the white matter is external, with masses of gray matter, that is, nerve cells, occupying a central position.

The cerebral hemisphere has been subdivided somewhat artificially into four lobes, shown in Fig. 48: the frontal (A), parietal (B), occipital (C), and temporal (D). All parts of the cortex do not have the same function, so gyri and sulci have been used to provide landmarks when we wish to describe an area of the cortex having a specific function.

Motor Cortex. One important landmark is the **central sulcus** (H). The gyrus immediately in front of the central sulcus is the **precentral gyrus** (I). This area of the cortex initiates voluntary movement in the opposite half of the body; the nerve fibers from the left precentral gyrus cross to the right side of the brain and spinal cord to reach those muscles on the right half of the body. For this reason it is called the **motor cortex.** What advantage, if any, this crossing, or **decussation,** provides is not known.

Sensation. The **postcentral gyrus** (J) is concerned with the recognition of ordinary sensations, such as pain and temperature, from the opposite side of the body, and is called the **sensory cortex.** Special sensations, such as the recognition of sound, are located at area F in the temporal lobe adjacent to the **lateral sulcus** (G).

Motor-Speech Area. In the left hemisphere in right-handed persons there is a motor-speech area (E), which coordinates the muscles used in speaking. Injury to this area does not produce paralysis in any muscle, but the individual has difficulty in saying words; this is called **motor aphasia.**

Visual Centers. On the medial surface of the right cerebral hemisphere the four lobes are indicated (Fig. 49). The **parieto-occipital sulcus**

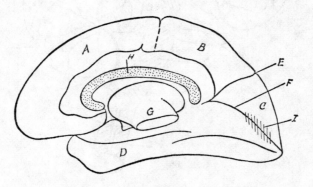

Fig. 49

is shown at E and the **calcarine sulcus** at F. The visual center of the cerebral cortex borders the posterior portion of the calcarine sulcus (I). Destruction of both visual centers results in total blindness.

Corpus Callosum. A large mass of nerve fibers (H) connects the two cerebral hemispheres. It has a characteristic shape when cut across and is called the **corpus callosum.** The cerebral hemispheres are connected to the remaining portion of the brain at the region (G) called the **midbrain.** In Fig. 50 the cerebral hemispheres have been separated by a sagittal cut through the corpus callosum (K), and the remaining parts of the brain have been cut in the same plane. The medial surface of the right cerebral hemisphere is shown at A with the cut surface of the corpus callosum (K).

Parts of Brain. Starting at the spinal cord (B) and moving upward, is the medulla oblongata (C). The foramen magnum (I) is the dividing line between the medulla and the spinal cord. Attached to the roof over the pons (D) is

the cerebellum (F). The midbrain (E) connects with the hemisphere.

Ventricles. Since the brain started in the embryo as nerve tissue surrounding a hollow tube, parts of the tube have dilated to form ventricles filled with cerebrospinal fluid. There is one such ventricle, called a **lateral ventricle,** in each cerebral hemisphere. Each lateral ventricle communicates via a small opening with the third ventricle (G), as shown by the arrow (see inset). Projecting from the roof of the third ventricle is the **pineal gland** (H). Its function in humans is as yet unknown. The third ventricle communicates with the fourth ventricle, (L), by a small channel, the **aqueduct** (M), that runs through the midbrain. Note that the central canal of the spinal cord also opens into the fourth ventricle. Cerebrospinal fluid leaves the ventricular system via three openings in the roof of the fourth ventricle, one of which is shown at J.

If a special liquid is injected into the ventricles which later hardens to form a cast of the actual spaces, the interconnections of the ventricular

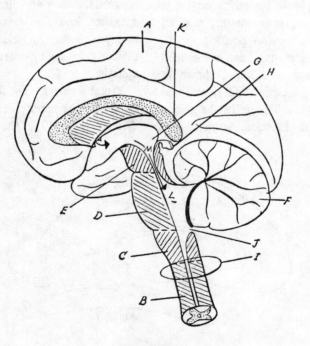

Fig. 50

system can be more readily seen (Fig. 50A). The right and left lateral ventricles (N) each communicating with the third ventricle (G) through small openings (O), while the third ventricle communicates in turn with the fourth ventricle (L) by means of the aqueduct (M).

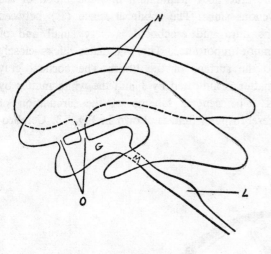

Fig. 50A

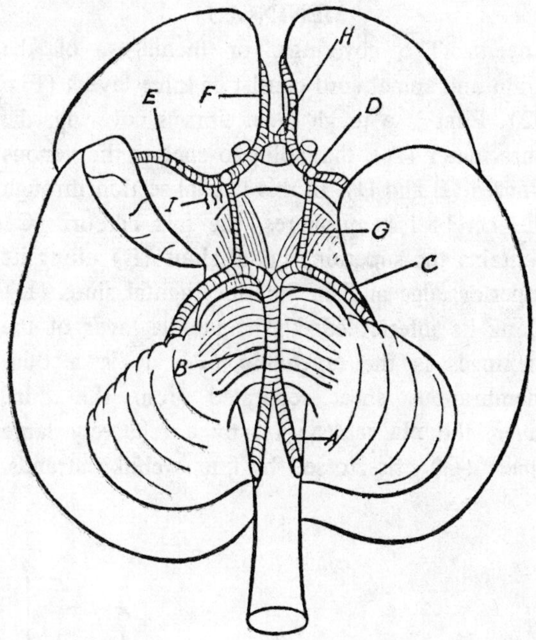

Fig. 51

Arteries. The major arteries supplying the brain are shown in Fig. 51. The two vertebral arteries (A) enter the cranial cavity through the foramen magnum to lie on the ventral surface of the medulla. They unite to form the **basilar artery** (B) on the pons. The basilar artery gives several arterial branches to the cerebellum (not shown), and ends by dividing into the right and left **posterior cerebral arteries** (C).

The other major artery to the brain is the **internal carotid,** the cut end of which can be seen at D. This artery divides into a **middle cerebral artery** (E) and an **anterior cerebral artery** (F). A small branch from the internal carotid artery, the **posterior communicating artery** (G), joins with the posterior cerebral branch (C) of the basilar artery. The anterior cerebral arteries are joined by a short **anterior communicating artery** (H). Observe how the arteries form a circle, known as the **circle of Willis.** Begin at the basilar (B), proceed to the posterior cerebral (C), to the posterior communicating (G),

to the internal carotid (D), to the anterior cerebral (F), across the anterior communicating (H) to the opposite anterior cerebral, and then in reverse order back to the basilar. These arteries lie in the subarachnoid space. For reasons that are not fully known, the vessels forming the circle frequently develop **aneurysms,** a dilation of a short segment of the artery. The aneurysm may press on adjacent nerves, causing paralysis, or the thin wall may rupture, with dire results.

The cerebral arteries described above supply the cortical areas of the hemispheres. In addition, small central branches arise from the arterial circle and penetrate the brain substance to supply deeper structures. One such important group is shown at I. These small branches supply blood to nuclei deep in the brain and intermingle with the nerve fibers coming from the motor cortex and going to the opposite side of the spinal cord. If these branches become blocked or if they rupture and produce hemorrhage in this area, the condition is called a "stroke."

MENINGES

Layers. The coverings, or **meninges**, of the brain and spinal cord consist of three layers (Fig. 52). First is a tough outer fibrous covering, the dura mater (A), that splits to enclose the venous sinuses (B and D). In this frontal section through the cerebral hemispheres, the falx cerebri (C) contains the superior sagittal sinus (B) along its superior edge and the inferior sagittal sinus (D) along its inferior edge. The middle layer of the meninges is the **arachnoid** (F). It is a thin, membranous sheet separated from the third layer, the **pia mater** (I), by a relatively large space (G) crisscrossed by fine, weblike strands.

This space is the **subarachnoid space** and contains cerebrospinal fluid.

Cerebrospinal Fluid Circulation. Tumorlike processes, of the arachnoid layer, called **arachnoid granulations** (H), project into a venous sinus (B). Cerebrospinal fluid is filtered through the arachnoid granulation into the blood of the venous sinus. The **subdural space** (E) between the dura and arachnoid is very small and of minor importance. The pia mater sticks closely to the surface of the brain. The cortical gray matter is indicated by J and the white matter by K. The general scheme of the circulation of cerebrospinal fluid is shown in Fig. 53. Cerebro-

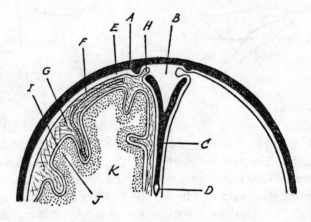

Fig. 52

spinal fluid is produced by a plexus of blood vessels in each of the four ventricles. The lateral ventricle (A) of the right hemisphere joins the third ventricle (C) through the **interventricular foramen** (B). Fluid from the third ventricle passes through the aqueduct (D) to the fourth ventricle (E), which also receives fluid from the central canal of the spinal cord (G). From the fourth ventricle the fluid reaches the subarachnoid space (H) through aperatures in its roof and sides (F). Cerebrospinal fluid in the subarachnoid space is filtered through the arachnoid granulation (I) to reach the bloodstream, that is, a venous sinus (J).

Functions. The functions of the cerebrospinal fluid are little known other than that it carries waste products from the brain to the bloodstream. It may have a protective function in that it forms a fluid cushion around the brain and spinal cord. The fluid is removed by inserting a needle in the lumbar region of the back to pierce the subarachnoid space surrounding the spinal cord, a procedure called a **spinal tap.** Blockage of the flow of cerebrospinal fluid from the ventricles into the subarachnoid space results in enlargement of the ventricle as the result of increased pressure of the fluid. This condition is known as **hydrocephalus.**

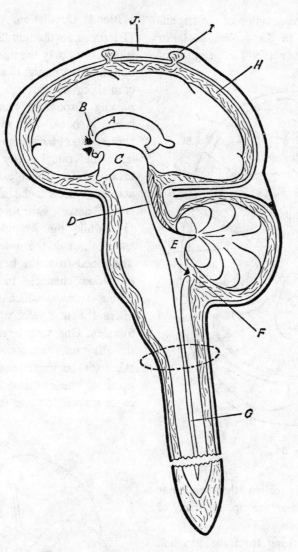

Fig. 53

CRANIAL NERVES

The nerves leaving the brain and the blood vessels going to the brain pass through openings, or **foramina,** in the floor of the skull (Fig. 54). The first cranial nerve, the **olfactory nerve,** arises in the nasal mucosa and passes to the olfactory bulbs, which rest on the cribriform plate (A) of the ethmoid bone. The second cranial nerve, the **optic nerve,** arises in the retina of the eyeball and passes through the optic canal (B) to reach the brain. The **oculomotor, trochlear,** and **abducens** nerves, the third, fourth, and sixth cranial nerves, respectively, are motor nerves arising in the brain. They reach the orbit by passing through the superior orbital fissure (C) to supply the muscles that move the eyeball. The fifth cranial nerve, the **trigeminal,** is largely a sensory nerve carrying sensations—pain and temperature, for example—from the skin of the face as well as underlying structures, such as mucosa of eyelid, nasal cavity, air sinuses, lips, and teeth. The ganglion of the trigeminal nerve corresponds to the typical posterior root ganglion of a spinal nerve. The ganglion of

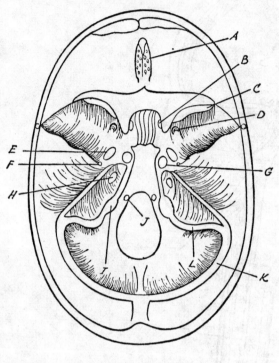

Fig. 54

the trigeminal nerve is called the **semilunar ganglion** and is found near the tip, or apex, of the petrous temporal bone.

The sensory fibers arising from the ganglion form three distinct branches. The ophthalmic division passes through the superior orbital fissure (C) to the orbit and the overlying structures of the orbit. The middle division, or maxillary, passes through the **foramen rotundum** (D) to supply the structures contained in the maxillary bone as well as the skin covering it. The mandibular division passes through the **foramen ovale** (E) to mandibular structures, including teeth, tongue, lower lip, and skin. The mandibular division also carries voluntary motor fibers that supply the muscles of mastication with the exception of the **buccinator,** which is supplied by the seventh cranial nerve, the **facial nerve.**

Blood Circulation. The **foramen spinosum** (F) transmits the middle meningeal artery, which supplies not only the dura mater but also the skull bones. The internal carotid artery enters the cranial cavity through the opening (G) after passing through a bony canal in the petrous temporal bone. The seventh and eighth cranial nerves, the facial and the auditory nerves, leave the cranial cavity via the **internal auditory meatus** (H). The **glossopharyngeal, vagus,** and **accessory** nerves, the ninth, tenth, and eleventh cranial nerves, exit through the **jugular foramen** (I), while the **hypoglossal,** or twelfth, nerve leaves through the **anterior condylar canal** (J). The blood from the brain is drained by a series of blood channels in the dura mater called venous sinuses, which meet to empty into the internal jugular vein, which begins at the jugular foramen. One such large channel can be seen in the diagram, the **transverse sinus** (K), which skirts the common boundary of the middle and posterior cranial fossae. It continues into the posterior cranial fossa as the **sigmoid sinus** (L) and

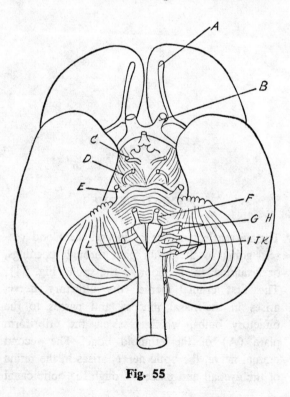

Fig. 55

empties into the beginning of the internal jugular vein.

Attachments. The attachment of the cranial nerves to the ventral surface of the brain is shown in Fig. 55 and is summarized in Table 1. Cranial nerves are customarily numbered with Roman numerals.

Table 1

Nerve	Location in Fig. 55	Attachment
I. Olfactory	A	Cerebrum
II. Optic	B	Cerebrum
III. Oculomotor	C	Midbrain
IV. Trochlear	D	Midbrain
V. Trigeminal	E	Pons
VI. Abducens	F	Junction of pons and medulla
VII. Facial	G	Junction of pons and medulla
VIII. Auditory	H	Junction of pons and medulla
IX. Glosso-pharyngeal	I	Medulla
X. Vagus	J	Medulla
XI. Accessory	K	Medulla
XII. Hypoglossal	L	Medulla

THE EYE

The bony socket of the eyeball, called the **orbit,** is cone-shaped. In the horizontal section shown in Fig. 56, looking into the right orbit from above, the medial wall (A) runs directly forward, and the lateral wall (B) makes an angle of 45 degrees with it. Both walls are of about equal length, two inches, but because of the inclination the lateral surface of the eyeball is exposed, and the surgical approach to the eyeball is usually from the lateral side. The **optic nerve** (C), carrying sensation from the **retina,** passes directly to the apex of the orbit on its way to the brain.

The eyeball, about one inch in diameter, projects forward beyond the anterior opening or base of the orbit, and the space behind the eyeball is filled with muscles, nerves, and fat. The sunken eyes of an emaciated person result from a decrease in orbital fat.

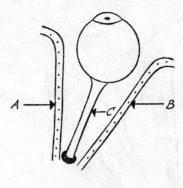

Fig. 56

Muscles. Movements of the eyeball are performed by six muscles arising from the apex of the orbit. In Fig. 57, four of these muscles are

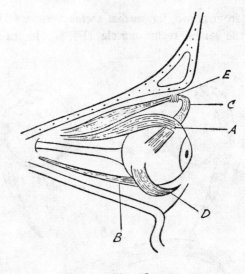

Fig. 57

seen in a side view, or sagittal section, through the orbit. The **superior rectus muscle** (A) and the **inferior rectus muscle** (B) are inserted on the upper and lower surfaces of the eyeball. The **superior oblique muscle** (C) runs through a fascial sling or pulley at E and bends back to be attached obliquely on the upper surface of the eyeball. The **inferior oblique** (D) is the shortest muscle. It arises from the floor of the front of the orbit and runs obliquely on the inferior surface of the eyeball. In Fig. 58, looking into the

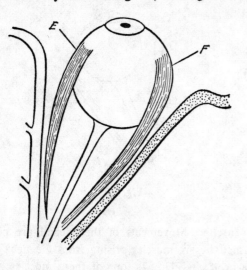

Fig. 58

orbit from above, the **medial rectus muscle** (E) and the **lateral rectus muscle** (F) are inserted

into their respective surfaces of the eyeball. The trochlear, or fourth cranial, nerve (*trochlea* means "pulley") is the motor nerve of the muscle with a pulley, the superior oblique. The abducens, or sixth cranial, nerve supplies the lateral rectus, while the oculomotor, or third cranial, nerve supplies all the other muscles of the eyeball.

Movements. In Fig. 59 the movements of the eyeball produced by the contraction of these muscles are shown. The right eyeball is shown throughout. In Fig. 59A the medial rectus (MR) **adducts** the eyeball by turning the pupil toward the nose. The lateral rectus (LR) **abducts** the eyeball by turning it in the opposite direction (Fig. 59B). **Elevation** of the eyeball (turning the pupil upward) is performed by two muscles, the superior rectus (SR) and the inferior oblique (IO), as Fig. 59C shows. The opposite movement, **depression,** or looking downward, is due to the superior oblique (SO) and the inferior rectus (IR), as shown in Fig. 59D. **Rotation** of the eyeball around its anteroposterior axis is shown in Figs. 59E and 59F. Medial rotation of the right eyeball (Fig. 59E), often called **in-**

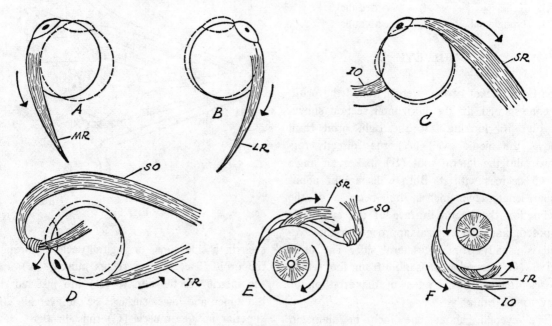

Fig. 59

torsion, results in a clockwise turning of the pupil; the superior oblique (SO) and superior rectus (SR) muscles are responsible. The opposite movement, lateral rotation or **extorsion** (Fig. 59F), is produced by the inferior oblique (IO) and the inferior rectus (IR), muscles.

Layers of Eyeball. For simplicity, each of the three layers of the eyeball is shown separately. In Fig. 60 the outer layer, or **sclera** (A) forms

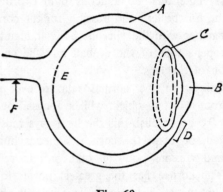

Fig. 60

the white of the eye. In children it is thin and appears to be blue because the deeper layer, the **choroid,** shows through. The anterior part of the sclera becomes clear and transparent to form the **cornea** (B). The cornea is about one-half inch in diameter. The **limbus** (D) is the circular line of junction between the cornea and the sclera and can be likened to the rim of the cornea. Degeneration of the sclera in the region of the limbus, often seen in elderly people, produces a gray ring called **arcus senilis.** Within the sclera, near the limbus, a channel (C), called the **canal of Schlemm** encircles the eyeball. This canal is concerned with the drainage of the fluid within the eyeball. The posterior wall of the sclera has sievelike openings (E) through which the fibers of the optic nerve pass. A part of the sclera forms a sheath (F) for the optic nerve. This sheath is continuous with the dura mater, the outer covering of the brain.

The middle coat, the choroid (Fig. 61), is

the vascular layer of the eyeball. A thickening of its anterior end forms the **ciliary body** (A), which contains the muscles necessary for ac-

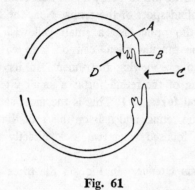

Fig. 61

commodation, to be discussed later. From the inner surface of the ciliary body project small fingerlike extensions, called **ciliary processes** (D), which contain many veins concerned with the production of aqueous fluid. The circular disc, or **iris** (B), with the hole in its center, or **pupil** (C), is the most anterior part of the choroid layer.

The lining of the eyeball is the **retina** (Fig. 62). It has an outer pigmented layer (A) and

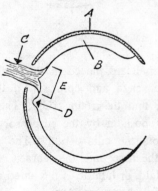

Fig. 62

an inner layer of nerve cells and fibers (B). Detachment of the retina, sometimes due to injury, is a separation of the pigment layer from the nerve cell layer. The nerve fibers (C) converge to leave the eyeball at its posterior pole (E) forming a compact cylinder of nerve fibers. This area of exit is called the **optic disc;** since it contains no nerve cells, it is often referred to as the "blind spot" of the retina. Near the lateral side of the optic disc a small yellowish spot, called the **macula lutea,** can be seen with the aid of an eye-viewing instrument. In its center a thinning of the retina forms a small pit called the **central fovea (D)**. This is the most sensitive part of the retina, and it is on this spot that light rays are focused when we look directly at an object.

Aqueous Humor. In Fig. 63 the three layers

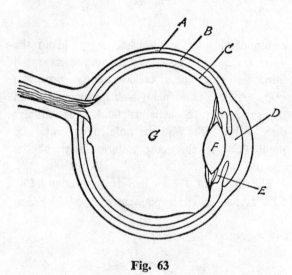

Fig. 63

of the eyeball are placed in their relative positions (the sclera and cornea [A], the choroid layer [B], and the retina [C]). The **lens** (F) is held in position by the suspensory ligaments arising from the ciliary body. The space (D) between the iris and the cornea, filled with aqueous fluid, or humor, is the **anterior chamber;** the **posterior chamber** (E) is the fluid-filled space between the lens and the iris. The ciliary processes in the posterior chamber produce the aqueous humor that flows into the anterior chamber through the pupil. The fluid is absorbed in the canal of Schlemm near the angle between the iris and the rim of the cornea. Excess production

of aqueous humor or interference with its absorption raises the pressure within the eyeball; if this condition, called **glaucoma,** is not treated promptly, it will result in permanent blindness. Pressure within the eyeball can be measured by an instrument that, when pressed on the surface of the anesthetized eye, records the external pressure required to indent the surface of the eyeball. The space G is filled with a gel-like substance called the **vitreous body,** a clear structureless material. The aqueous humor provides nutrition for the lens and possibly for the cornea, and together with the vitreous body it maintains the proper shape of the eyeball and the correct position of the lens.

Focusing. In the normal relaxed eye, light rays from a distant object will be focused on the retina. To accomplish this the light rays must be refracted, or bent; the cornea, aqueous humor, lens, and vitreous body all take part in refraction. The cornea has the greatest power to refract light waves, but only the lens has the ability to alter its refracting power. The extremely divergent light rays from an object ten inches in front of the eye would be focused behind the retina in the relaxed eye, and as a result vision would be indistinct. Fortunately, an automatic movement takes place within the eyeball that increases the refractive power of the lens so that distinct vision is possible. This mechanism for near vision is called **accommodation.**

The lens in the normal young person is rather soft, and its shape is determined in part by the pressures and tensions exerted on it. In the relaxed eye, tension is exerted on the lens by the suspensory ligaments attached to its periphery (Fig. 64). Contraction of the radiating fibers (A) will pull the ciliary body forward, and contraction of the circular fibers (B) will bring the ciliary body closer to the edge of the lens, somewhat like the pull on a purse string. Since the distance between the points of the attachment of the suspensory ligaments (C) is shortened, their pull on the lens is decreased and the lens changes shape. The most conspicuous change in the lens is a greater convexity, or forward bulging, of its anterior surface and a resulting increase in its refractive power. Light rays from a near object can now be focused on the retina.

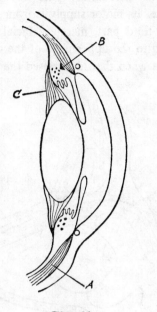

Fig. 64

Light Reflex. The muscles of accommodation receive their motor supply from parasympathetic nerve fibers traveling in the oculomotor nerve. If we pass a light in front of the eye of a person seated in a darkened room, we will observe that the pupil gets smaller, or **constricts.** This response of the pupil to light is called the **light reflex.** The iris, like a diaphragm, can change its width due to the actions of two muscles on its posterior surface (Fig. 65). The circular fibers (B) near

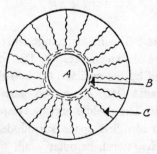

Fig. 65

the periphery of the pupil (A) contract and so shorten to constrict the pupil. This **sphincter muscle** is supplied by parasympathetic nerves. **Dilation** or enlarging of the pupil is due to the relaxation of this muscle. The radiating muscle fibers (C) can also dilate the pupil and are supplied by the sympathetic nerves.

Eyelids. The **orbicularis oculi muscle** functions to close the eylids (Fig. 66). The more

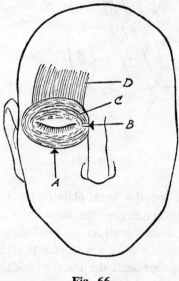

Fig. 66

peripheral part of the muscle surrounds the orbit (A), and the central part (C) is found within the eyelids. The muscle is attached to the medial angle of the eye (B). The **frontalis muscle,** in the eyebrow (D) helps to raise the upper lid. These muscles are supplied by the facial nerve. If the facial nerve is paralyzed, from Bell's palsy or following a stroke, the eyelids on the affected side cannot be closed.

The structures found within the upper eyelid are shown in a sagittal section in Fig. 67. The anterior surface is covered by skin (A) and the inner side is linked with an epithelium called the **conjunctiva** (B). The conjunctiva is reflected onto and covers the visible part of the eyeball. The space between the lid and the eyeball is called the **conjunctival sac** (C). Secretions from the **lacrimal gland** (D) enter the conjunctival sac through several ducts and form the tears. The rigidity of the lids is caused by cartilaginous

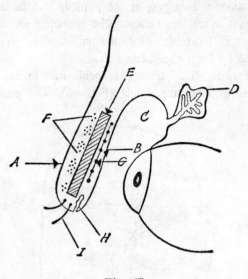

Fig. 67

plates; the **superior tarsal plate** (E) is the larger one. Lying anterior to the tarsus, the cut ends of the orbicularis oculi muscle are indicated (F). Posterior to the plate are the **tarsal glands** (G) which, together with the tears, provide lubrication.

At the edge of the lid are found the **ciliary glands** (H), which are opened by small ducts near the eyelashes (I). A **sty** is an inflamed ciliary gland.

When looking for a foreign object in the conjunctival sac, place a small thin stick, such as a wooden match, against the upper lid just below the eyebrow and parallel with it. Using your other hand, grasp a few eyelashes and raise the lid. The bending and eversion of the lid takes place above the upper limit of the tarsal plate. The normal rigidity of the plate will keep the lid everted.

The chief muscle responsible for raising the upper lid originates in the orbit and is shown in a sagittal section (Fig. 68). The **levator palpebrae** (A), literally "elevator of the lid," lies above the superior rectus and is divided into three parts. The major part (B) is in the skin of the upper lid and is supplied by the oculomotor nerve. The levator palpebrae muscle is somewhat unusual, because part of it is nonstriated muscle. This involuntary part (C) goes to the tarsal plate. Its motor supply is from sympathetic nerves. A third part, merely a facial band (D), is attached to the upper edge of the conjunctival sac so that when the lid is raised the conjunctiva

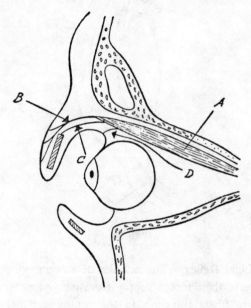

Fig. 68

is not left behind. As a consequence of the dual nerve supply of this muscle, dropping of the upper lid, called **ptosis**, can be due to paralysis of either the oculomotor or the sympathetic nerves.

THE EAR

Components. The major components of the human ear are shown in Fig. 69. The sound waves are funneled by the **auricle** (A) into the **external auditory canal** (B). These two parts form the **external ear.** The **middle ear,** or **tympanic cavity** (C), is an air-filled space surrounded by bone except for its lateral, or outer, wall, which is the **tympanic membrane.** Thus, the boundary between the external and middle ear is the tympanic membrane, or **eardrum.** The middle ear cavity contains the earbones, or **ossicles,** which transmit and amplify the movements of the tympanic membrane to the **inner ear** (D).

Inner Ear. The inner ear is an elaborate system of spaces connected by passageways in very dense bone. Because of its resemblance to a labyrinth, it is also known by that term. The inner ear contains specialized structures for transposing sound waves to nerve impulses, which are a form of electrical energy. Other structures record motion of the head as well as its position in respect to gravity. Nerve fibers from these

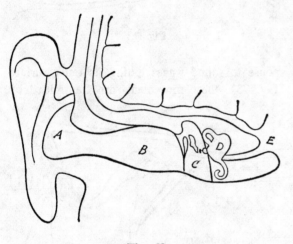

Fig. 69

structures leave the inner ear through a bony tube (E), called the **internal auditory canal,** on their way to the brain.

Middle Ear. The middle ear cavity can be thought of as a room (Fig. 70) that has four walls, a roof, and a floor. The lateral, or outer, wall formed by the tympanic membrane has been removed so you can see inside. The anterior wall to your right contains an opening (A) that is one end of a tube that connects with the **nasopharynx** at its other end. This is called the auditory, or **Eustachian,** tube. If normal hearing is to be maintained, the pressure of the atmosphere on both sides of the tympanic mem-

brane must be equal. Air entering the nose to the nasopharynx and moving along the auditory tube to the middle ear cavity will exert the same

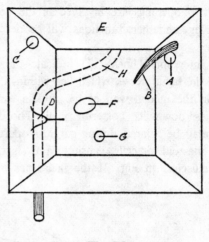

Fig. 70

pressure on the inside of the tympanic membrane as air reaching the outer side of the membrane through the external auditory canal. The **tensor tympani muscle** (B) arises from the anterior wall and is attached to an ear bone, the **malleus.** Since the malleus is attached in turn to the tympanic membrane, the tensor tympani muscle, as its name suggests, maintains the correct tension of the eardrum.

On the posterior wall an opening (C) can be seen; it is the **aditus,** or doorway, to the **mastoid air cells.** As a result of this opening an infection of the middle ear cavity frequently involves the mastoid air cells. Inferior to the additus there is a bony projection (D) called the **pyramid,** from which arises a small muscle called the **stapedius.** This muscle is attached to the smallest of the ear ossicles, the **stapes.**

The bony roof (E) of the middle ear cavity is thin, and a part of the brain rests on its upper surface. Because of this juxtaposition, an infection of the middle ear cavity can sometimes cause an abscess in the brain.

The medial, or inner, wall has two openings, or windows. The **oval window** (F), so-called because of its shape, opens into the internal ear. This window is covered by the footpiece of the stapes. The **round window** (G) also opens into the internal ear but is covered by a membrane. These two openings are involved in the process of hearing, and their functions will be described later.

The interrupted lines (H) indicate the position of the **facial nerve,** which lies within a bony canal in the medial wall of the middle ear and continues down its posterior wall. The facial nerve may be injured during surgical operations on the mastoid air cells, causing paralysis of the facial muscles. In Fig. 71 the facial nerve (A)

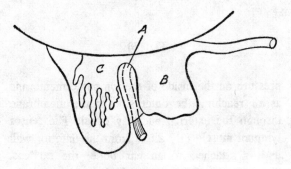

Fig. 71

is shown within the bony wall separating the middle ear cavity (B) from the mastoid air cells (C). Since all infected mastoid cells must be removed, it will be seen that the nerve is in an exceedingly vulnerable position.

The function of the ear ossicles is explained in Fig. 72. The malleus, or hammer (A), is attached to the tympanic membrane (B). The malleus articulates, or forms a joint, with the **incus,** or anvil (C). The incus in turn articulates with the stapes, or stirrup (D). The footpiece of the stapes fits against the oval window (E), and the fluid in the internal ear is in contact with that part of the stapes. A circular or **annular** ligament (F) attaches the footpiece of the stapes to the edge of the oval window so that the fluid cannot escape into the middle ear

cavity. This ligament is lax enough to allow movement of the stapes.

As the tympanic membrane vibrates in re-

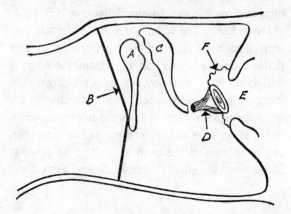

Fig. 72

sponse to sound waves hitting its outer surface (Fig. 73), these movements are transmitted by

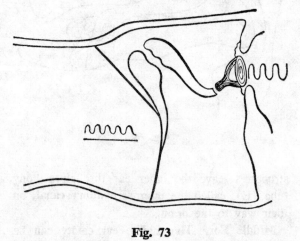

Fig. 73

the ossicles to the fluid of the internal ear. Because the ossicles act as a complex system of levers, the amplitude of movement of the footpiece of the stapes is about twenty times that of the tympanic membrane. The stapes does not move in a simple medio-lateral direction, but moves as a foot tapping while keeping the heel in contact with the floor. Excessive movement of the ossicles due to a loud noise is prevented by contraction of the **stapedius** and tensor tympani muscles.

Mechanical Impairment of Hearing. We should now consider those conditions that interfere mechanically with the transmission of sound to the internal ear. The most common example is the temporary impairment of hearing we observe when taking off or landing in an airplane. In Fig. 74 the outer surface of the tympanic

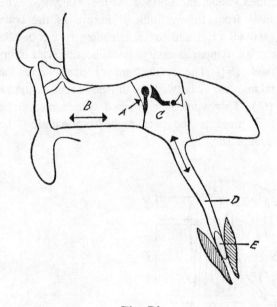

Fig. 74

membrane (A) is in contact with the atmosphere through the external auditory canal (B). The inner surface of the membrane is influenced by the air pressure in the middle ear cavity (C). The auditory tube (D) allows the middle ear cavity to communicate with the atmosphere via the nose so that the air pressure on both sides of the eardrum is equal. The nasopharyngeal opening (E) of the auditory tube is surrounded by muscles that normally squeeze the inlet shut. These muscles are used in swallowing and yawning, and when they contract, the inlet of the auditory tube gapes open.

During the ascent of the airplane the atmospheric pressure on the outside of the tympanic

membrane decreases, while the pressure within the middle ear cavity remains the same because the auditory tube has remained closed. The membrane is now pushed laterally toward the external auditory canal by the relatively higher pressure in the middle ear cavity. Vibration of the now taut membrane is limited, with the result that hearing is less acute. The act of swallowing or yawning will open the inlet of the auditory tube, the pressures on both sides of the membrane rapidly equalize, and hearing is restored. During descent, since the relative pressures are reversed, the tympanic membrane is pushed medially into the middle ear cavity. The treatment, however, is the same.

Fluid in the middle ear cavity will interfere mechanically with the transmission of vibrations from the tympanic membrane to the internal ear. The fluid may be caused by an infection, called **otitis media,** or by injury. Blockage of the opening of the auditory tube by the **adenoids,** a condition found in children, may also result in hearing loss. The adenoids are masses of lymphoid tissue in the nasopharynx, and lymphoid tissue grows most rapidly in childhood. After childhood all lymphoid tissue gradually shrinks. The tympanic membrane is pushed toward the middle ear cavity, indicating lower atmospheric pressure in the cavity. The cells of the epithelial lining of the middle ear cavity absorb oxygen from the air within that cavity, and since the source of air from the outside is blocked by the adenoids, the atmospheric pressure falls.

The overgrowth of bone around the oval window often involves the ligament holding the foot-piece of the stapes in position. Movement of the stapes may now be limited or prevented. This condition is called **otosclerosis.** Hearing is often restored either by surgically freeing the stapes or by drilling a new oval window through the bone into the internal ear. The new window is covered by a skin membrane so that inner ear fluid does not leak into the middle ear cavity. A connection is then made between the ossicles and the newly fashioned membrane.

The interconnected spaces hollowed out of the dense bone are indicated by the vertical parallel lines in Fig. 75. The space S is the **semicircular canal.** Although only one canal is shown here, there are two others lying in different planes. These canals provide information about the movement or motion of the head. The space V is the vestibule concerned with the position of the head with respect to gravity. The remaining space, C, is the cochlea, which contains the hearing organ.

Membrane Sacs. Suspended within each of these bony spaces is a corresponding membrane sac. Within the bony semicircular canal the membranous sac is called the **semicircular duct.** However, in the bony vestibule there are two interconnecting sacs. The larger one is called the **utricle** (A), and the smaller one is the **saccule** (B). The elongated sac in the cochlea is called the **cochlear duct** (D). The membranous sacs are filled with fluid called **endolymph,** indicated by the clear areas.

Perilymph. The fluid filling the space between the membranous sacs and the bony wall is called the **perilymph** and is indicated by the stippled area. In the cochlea the perilymph is contained in two channels (E and F) separated by the cochlear duct (D). Because they resemble the winding of a staircase, these two channels are called **scalae,** the Latin word for "stairway." The scala from the vestibule is known as the **scala vestibuli** (E), and the one leading to the middle ear, or tympanic cavity, is called the **scala tympani** (F). The scala tympani opens into the middle ear cavity through the round window (G). This window is covered with a membrane that prevents the perilymph from entering the middle ear.

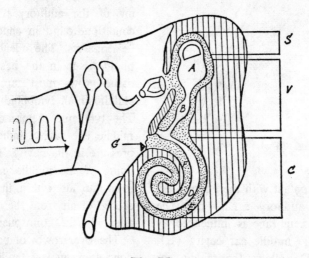

Fig. 75

In Fig. 76 a magnified section of a part of the cochlea is shown. The scala vestibuli (E) and the scala tympani (F) are filled with perilymph. The cochlear duct (D), filled with endolymph, contains the hearing organ, or **organ of Corti** (H). This consists of tall epithelial cells from which project tiny hairs, which are embedded in the overhanging roof of gelatinous material. Passing from the hearing organ are neurons (N) that go to the cortex of the brain. It is believed that movement of the hairs results in nerve impulses that are interpreted by the brain as sound. How one is able to distinguish different sounds is not yet clearly understood.

Sound Waves. Follow the path of the sound wave in Fig. 75. The wave enters the external auditory canal, producing vibration of the tympanic membrane. Movement of the membrane is transferred and amplified by the ossicles through

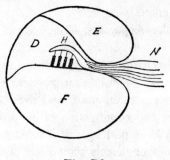

Fig. 76

the oval window to the perilymph. Movement in the perilymph travels in the scala vestibuli and may be transferred through the membranous walls of the cochlear duct to the endolymph. Movement in the endolymph produces movement of the hairs of the organ of Corti and the resulting nerve impulses travel to the brain.

Since fluid is almost incompressible, no movement could occur in the perilymph unless expansion is permitted at the other end of the perilymph channel. Note that the scala vestibuli is continuous with the scala tympani at the blind tip of the cochlear duct. It is the membrane covering the round window that allows temporary expansion of the perilymph. As the stapes moves to compress the perilymph at the oval window, there is a corresponding bulging of the round window membrane into the middle ear cavity.

Crista. The delicate structure called the **crista,** which detects movement of the head, is shown in Fig. 77. It is found in the bulbous portion of the membranous semicircular duct called the **ampulla** (A). It is composed of tall epithelial cells (B) from which emerge hairs (C). The hairs are embedded in the gelatinous mass (D). The flow of endolymph in the direction indicated by the arrow produces increased pressure in the ampulla. As a result of the increased pressure the crista sends the message to the brain.

The flow of endolymph within the duct may be likened to the movement of water in a round pan that is rotated. The apparent flow of the

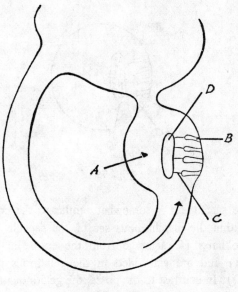

Fig. 77

water is opposite to the movement of the pan. Thus, when the head is rotated to the right (Fig. 78), pressure is increased in the right

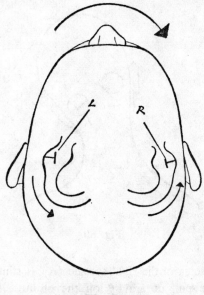

Fig. 78

ampulla (R) and decreased in the left (L).

Macula. Changes in the position of the head are detected by a structure called the macula (Fig. 79), which is found in the utricle and in

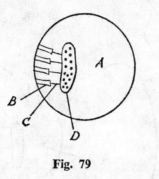

Fig. 79

the saccule. It is somewhat similar to the crista. Within the membranous sac of the saccule (A) the hairs (C) emerge from the epithelial cells (B) and are embedded in the gelatinous mass (D). In contrast to the crista, the gelatinous mass of the macula contains particles of the inorganic material called **otoliths** (literally, "ear stones"). The pull of gravity on the otoliths exerts tension on the hairs, which stimulates the nerves of the macula to transmit the information to the brain. In Fig. 80 the head is bent to the right, and

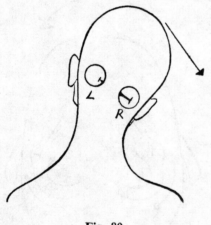

Fig. 80

the macula of the right saccule (R) is stimulated by the pull of gravity on the otoliths. In the left saccule (L) there is no pull on the hairs

as the otoliths rest against the epithelial cells. The macula in the utricle detects changes in the position of the head as it is bent forward or backward, but the mechanism is essentially the same as in the saccule.

Vestibular Apparatus. There is an extremely complex system of interconnections within the brain between the vestibular apparatus and those brain centers responsible for muscular movements of the limbs, trunk, head, and eyes. Our sense of balance and coordination of body movements depend in large part upon it. People with vestibular injuries or defects should not dive or swim under water because they are unable to judge the direction of the water surface. Excessive stimulation of the vestibuli can produce changes in blood pressure, heart rate, respiration, and movement of the gastrointestinal tract, resulting in the condition known as **motion sickness.** We have seen that gravity is a constant and potent factor in the proper functioning of the vestibular apparatus. An environment without gravity was so unusual until a few years ago that there was little understanding of its effects.

THE NOSE AND ASSOCIATED AIR SINUSES

The nose is a rectangular passageway divided into right and left nasal cavities by a **median septum.** Fig. 81, a sagittal section through the head slightly to the right of the septum, illustrates the different regions of the nasal cavity. Air enters through the **nostril** (A) and transverses the nasal cavity to enter the **pharynx** (B). The opening from each nasal cavity into the pharynx, indicated by the vertical dotted line, is called the **posterior choana.** The part of the cavity indicated at C is called the **vestibule** and is lined with skin. Hairs found in this area serve to trap dust particles mechanically. The **olfactory region** (D) is found in the upper one-third of the cavity. The remainder is known as the **respiratory area.** The olfactory and respiratory areas are covered by mucous membrane, which in the respiratory area is additionally provided with cilia. The cilia, by beating in a coordinated fashion, mechanically move the mucus secreted by the membrane posterior to the choanae and thence to the pharynx to be swallowed. Special nerve cells concerned with smell are found in the mucous membrane of the olfactory area. These cells send filaments through the sievelike cribriform plate of the ethmoid bone to the **olfactory bulb** (E) which carries the information to the brain.

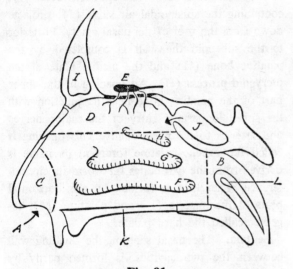

Fig. 81

line. The width at the level of the floor in the adult is approximately one-half inch, but near the roof it is a mere slit because the ethmoid bone bulges medially into the cavity. The length of a nasal cavity is about two inches, as is the height in the middle of the cavity.

Air Sinuses and Nasal Cavities. The relationship between the air sinuses and the nasal cavities is shown schematically in Fig. 82. The nasal

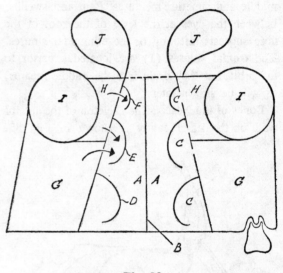

Fig. 82

Nasal Cavity. Projecting into each nasal cavity from the lateral wall are three shelflike structures called the **turbinates** (F, G, and H) that increase the surface area of the respiratory region. At the anterior part of the roof of the nose the **frontal air sinus** (I) is found. The posterior part of the roof is pushed into the nasal cavity by the bulging nonpaired sphenoid bone, which contains the **sphenoidal air sinus** (J). The floor of the nasal cavity is the **hard palate,** which also forms the roof of the mouth (K). In line with the inferior, or lowest, turbinate (F) is the opening of the auditory, or Eustachian, tube (L) in the nasal part of the pharynx. The auditory tube connects the pharynx with the middle ear and mastoid cells.

The nasal cavity is somewhat triangular in out-

cavities (A) are separated by the septum (B). The perforated roof of each nasal cavity permits passage of the filaments of the olfactory nerve. Projecting from the lateral wall of each cavity are the turbinates, named according to their relative position to each other: the highest or superior (F), middle (E), and inferior (D) turbinates. The space covered by a turbinate is called a **meatus** (C).

The largest air sinus related to the nose is the **maxillary** (G), which drains into the middle meatus. Because the opening of the maxillary sinus is high up on its medial wall, gravity does not help much in draining its secretion of mucus, or pus, if it becomes infected. Normally the beating of its ciliated epithelial lining will continuously move its mucus into the nasal cavity, but if the cilia are destroyed by infection, a surgical opening is made into the inferior meatus so that gravity will aid in drainage. The floor of the sinus is related to the molar teeth; thus, a sinus infection may result from an infected tooth, or a sinus infection may produce a toothache. The bony roof of the maxillary sinus is also the floor of the **orbit** (I).

Situated medial to each orbit are the paired **ethmoid air sinuses** (H), which drain into the middle and superior meatuses. Pain and swelling between the eyes at the level of the root of the nose suggests infection of the ethmoidal sinuses. The **frontal sinuses** (J) are located superior to the orbit, but they drain into the middle meatus, as will be shown later.

Bones of the Nose. The skeleton of the lateral wall of the nasal cavity is shown in Fig. 83.

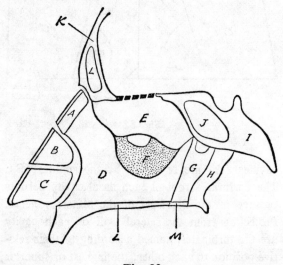

Fig. 83

The external nose is formed by paired nasal bones (A), which are attached to the frontal bone (K). This junction is known as the **root** of the nose. The inferior half of the external nose is formed by cartilages (B and C) that allow some mobility. Note that the skin is mobile over the nasal bones but is firmly attached to the cartilaginous part. A pimple in the skin over the cartilaginous part is therefore much more

painful, because a swelling in the already taut skin exerts greater pressure on the nerve endings.

The **ethmoid bone** (E) forms the upper part of the wall; on the other side of this thin bone are the honeycomblike air cells of its sinus. The lower part of the wall is formed by the **maxillary bone** (D), which separates the maxillary sinus from the nasal cavity. A considerable gap between the two bones is filled by a membrane (F), through which the opening into the maxillary sinus can be seen. The **sphenoid bone** (I), containing the sphenoidal air sinus (J), projects down from the roof of the nasal cavity. Posterior to the sphenoid the wall is completed by the **palatine bone** (G) and the medial plate of the **pterygoid process** (H). An opening in the upper part of the palatine bone near its junction with the sphenoid permits entry of the major nerves and arteries to the nasal cavity. This opening is called the **spheno-palatine foramen;** its name is derived from the two bones that form its circumference. The floor is formed by the horizontal parts of the maxilla (L) and palatine (M), together called the hard palate.

Septum. The nasal septum, the dividing wall between the two cavities, is formed partly by bone and partly by cartilage (see Fig. 84). The

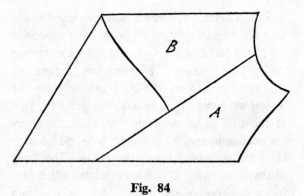

Fig. 84

vomer bone (A) forms the most posterior part of the septum, and the vertical plate of the ethmoid (B) projects from the roof. Cartilage anterior to the vomer and ethmoid completes the septum. **Deviation of the septum** is a bulging of the septum into either the right or left nasal cavities, which results in constriction or blocking of the air passage. It would appear that the height of the septum is sometimes too great to fit into the space provided. While injury to the nose during childhood may result in a deviated septum, there are other causes, largely unknown. The upper part of the septum is derived from a bone (ethmoid) whose growth is influenced by the growth of the brain; the lower part of the septum is influenced by the growth of the face bones. Since growth and development of the face and cranial skeleton take place at neither the same rate nor the same time, unequal growth of these components of the septum may produce septal deviation.

Blood Supply. The posterior two-thirds of the walls, roof, and floor of each nasal cavity is supplied by nerves and blood vessels that enter through the spheno-palatine foramen. The sensory nerves are branches of the maxillary division of the trigeminal. The arteries are branches of the internal maxillary. Bleeding from the posterior part of the nose arises from these vessels or their corresponding veins. The arteries run like horizontal pipes in the mucous membrane covering the turbinates, where they function to warm the inspired (breathed-in) air. Infection in the nasal cavity causes dilation of the blood vessels, which produces swelling of the mucous membrane, which in turn blocks the air passage. When substances containing adrenalin are applied to the membrane, the blood vessels will constrict. The anterior third of the nasal cavity gets its blood supply from the ophthalmic artery and its nerves from the ophthalmic division of the trigeminal nerve. The more frequent site of nosebleeds is in the inferior part of the septum near the nostril (A in Fig. 85). This area is supplied by a branch of an artery that runs in the upper lip, the **superior labial** artery (B). Bleeding may be controlled by placing cotton (C) between the lip

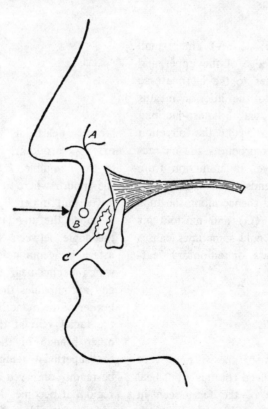

and the teeth and pressing the lip against the cotton.

Mucous Membrane. Since the mucous membrane of the nasal cavity is contiguous with that lining the air sinuses, infection or inflammation resulting from a "cold" involves all cavities.

Two other areas are similarly involved because of contiguous mucous membranes (see Fig. 86).

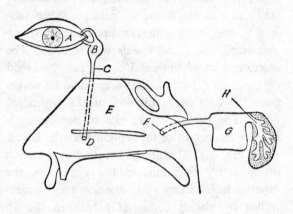

Fig. 86

Tears in the conjunctival sac (A) enter small canals that open on the edge of the upper and lower lids. The fluid passes to the lacrimal sac (B) and then drains into the inferior meatus (D) of the nasal cavity via the **naso-lacrimal duct** (C). Infection spreading in the direction opposite the flow of tears produces the characteristic redness of the eye. Inflammation from the nasal cavity (E) extends to the opening of the auditory tube (F) and thence along the tube to the middle ear cavity (G) and mastoid air cells (H). That is why a cold sometimes causes a clicking sound in the ears, or temporary deafness.

FACE

Blood Supply. The skin of the face is very vascular, so that wounds bleed copiously but heal rapidly. The chief artery of the face, seen in Fig. 87, is the **facial artery** (A), which is rel-

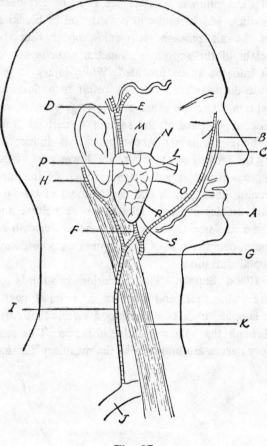

Fig. 87

atively superficial. The artery arises from the external carotid artery inferior to the lower border of the mandible. It travels toward the angle of the mouth where it gives a branch to both the lower and upper lips. Superior to this point it is called the **angular artery** (B), since it lies in the angle between the nose and the cheek. The artery is accompanied by the **angular vein** (C), which begins near the medial angle of the eye and accompanies the facial artery to the inferior border of the mandible. The angular vein becomes the **facial vein** at the angle of the mouth. Another branch of the external carotid artery is the **superficial temporal artery** (D), which can be readily observed and palpated in the "temple" region. It becomes increasingly tortuous with advancing age. The artery is accompanied by a vein

(E) of the same name. As this vein enters the substance of the **parotid gland** (P) it becomes the **retromandibular vein** (F), which divides into two branches. One branch joins the facial vein to form the **common facial vein** (G), which drains into the **internal jugular vein** lying deep to the sternomastoid muscle (K). The other branch of the retromandibular vein joins with the **posterior auricular vein** (H) to form the **external jugular vein** (I). The external jugular vein descends superficial to the sternomastoid muscle, where it empties into the **subclavian vein** (J) just above the clavicle.

Infections of the face are potentially dangerous for two reasons. The angular vein anastomoses with veins in the orbit which in turn drain into intracranial venous sinuses. Secondly, the veins of the face lack valves, so blood may drain into the cranial cavity as well as into the neck veins. If an infection of the face involves the facial vein, a clot, or **thrombus,** containing bacteria may form within the vein. Squeezing of the infected area may break off a small part of the thrombus, which could then be carried into the intracranial venous sinuses. An infected blood clot within the venous sinuses enlarges to produce blockage and pressure effects which are most serious. The area of the face between the angle of the mouth and the nose is known as the "danger area of the face."

Parotid Gland. Compressed between the ramus of the mandible and the external auditory meatus is the parotid gland (P). It extends from the level of the zygomatic arch to the angle of the mandible and is surrounded by dense fascia, which makes palpation of the gland difficult. Part of the gland lies superficial to the mandibular ramus. The **parotid duct** (L), carrying secretions from the gland, runs toward the cheek to pierce the **buccinator muscle** and open in the mouth at the level of the upper second molar tooth. The facial nerve, after leaving the stylomastoid foramen, enters the substance of the parotid gland on its way to supply the muscles of facial expression. Within the gland the facial

nerve breaks up into five branches, **temporal** (M), **zygomatic** (N), **buccal** (O), **mandibular** (R), and **cervical** (S). Tumors of the parotid gland, or even infections such as mumps, may produce sufficient pressure on the facial nerve to cause paralysis of the facial muscles.

Facial Nerves. The nerves carrying sensation from the skin of the face travel with the three chief divisions of the trigeminal nerve, seen in Fig. 88. The area indicated by A is supplied by the

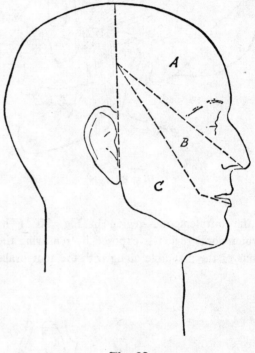

Fig. 88

ophthalmic division, Area B by the maxillary division, and Area C by the mandibular division of the trigeminal nerve.

Muscles of Mastication. There are five chief muscles of mastication. Two readily palpable muscles, shown in Fig. 89, are the **temporalis** (A) and the **masseter** (B). The temporalis muscle arises from the temporal fossa and inserts on the coronoid process and anterior border of the mandibular ramus. The masseter muscle arises from the zygomatic arch and inserts on the lateral surface of the mandibular ramus. Both muscles produce **occlusion** or closure of the jaws. Palpate them while "clenching" the jaws.

Two other muscles of mastication are seen

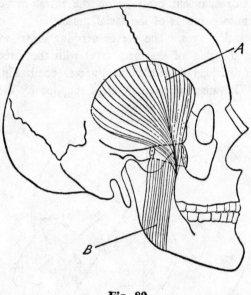

Fig. 89

in the infratemporal region in Fig. 90. (The infratemporal region is exposed by removing the ramus of the mandible along with the temporalis

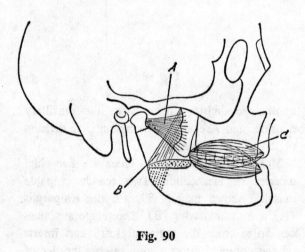

Fig. 90

and masseter muscles. The head and neck of the mandible are all that remain of the ramus.)

The more external **lateral pterygoid muscle** (A) arises from the pterygoid process and inserts on the mandibular neck and adjacent capsule of the jaw joint. This muscle depresses the jaw, as in opening the mouth, and also protrudes the jaw. The deeper **medial pterygoid muscle** (B) arises from the pterygoid process and inserts on the medial surface of the angle of the mandible. In concert with the temporalis and masseter muscles it also closes the jaws. The remaining muscle of mastication is the **buccinator** (C), which arises from the pterygoid process and medial surface of the mandible. The muscle fibers run forward toward the angle of the mouth where the fibers intertwine with the muscle fibers of the lips, the **orbicularis oris.** The buccinator muscle aids during mastication by keeping the food between the teeth. All the muscles of mastication, with the exception of the buccinator, are supplied by the motor component of the mandibular division of the trigeminal nerve. The motor supply to the buccinator muscle is from the facial nerve.

Blood Supply to Muscles of Mastication. The vessels and nerves related to the pterygoid muscles in the infratemporal region are shown in Fig. 91 in which the ramus of the mandible has been removed. The lateral pterygoid muscle is inserted into the neck of the mandible, while the deeper medial pterygoid muscle is inserted on the medial surface of the angle of the mandible. The external carotid artery (A) ends by dividing into the internal maxillary artery (B) and the superficial temporal artery (C). The internal maxillary enters the infratemporal region by passing medial to the ramus of the mandible. Here it gives off the **inferior alveolar artery** (D), which supplies the lower jaw and teeth. The **middle meningeal artery** (E) is an important branch of the internal maxillary. The middle meningeal artery passes superiorly medial to the lateral pterygoid muscle to enter the cranial cavity via the foramen spinosum. It supplies blood to the dura mater and bones of the cranium. The internal maxillary artery continues anteriorly

(F) to enter bony channels which enable smaller branches of the artery to supply the nasal cavity, the hard palate, the maxillary sinus, and the upper teeth.

The mandibular division of the trigeminal nerve leaves the cranial cavity through the foramen ovale; two branches of the division can be seen emerging between the pterygoid muscles.

The **inferior alveolar nerve** (G) accompanies the artery of the same name (D) into a bony canal of the mandible to carry sensation from the teeth back to the brain. It is this nerve which is "blocked" by a local anesthetic during dental repairs on the lower teeth. The remaining nerve is the **lingual** (H), which travels anteriorly to reach the tongue.

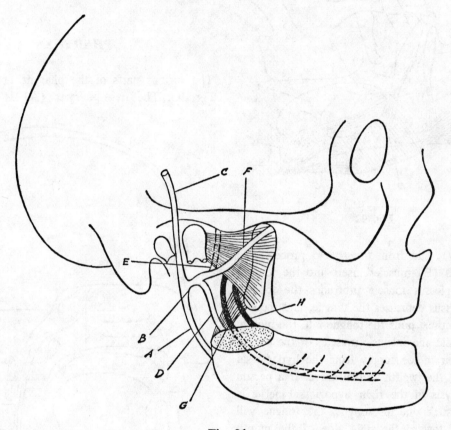

Fig. 91

TONGUE

The tongue lies in the floor of the mouth and consists of muscle covered with mucous membrane. The tongue is shown in Fig. 92. It functions in chewing, swallowing, speaking, and tasting. The dorsal surface of the tongue is subdivided by a V-shaped groove into two parts: the anterior part (A), which comprises two-thirds of

the dorsal surface, and the posterior part (B). The muscles of the tongue consist of two types. Those muscles found entirely within the tongue are called the **intrinsic muscles** and produce changes in the shape of the tongue. The **extrinsic muscles** arise from bones outside the tongue and are able to move the tongue as well as change its shape. The intrinsic muscles are not shown, but they consist of bundles of fibers running longitudinally, vertically, and transversely. There are three extrinsic muscles on each side of the tongue. From the mandible (J) the **genioglossus** (C), from the hyoid bone, the **hyo-**

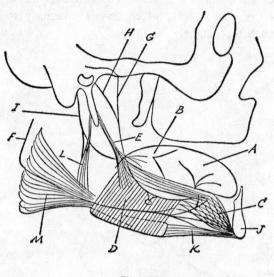

Fig. 92

glossus (D), and from the styloid process, the **styloglossus** (E) muscles insert into the tongue. The genioglossus muscle protrudes the tongue, the hyoglossus depresses the tongue, and the styloglossus muscle pulls the tongue into the mouth. The extrinsic and intrinsic muscles of the tongue receive their motor supply from the hypoglossal nerve (F), the twelfth cranial nerve. If a person with paralysis of the right hypoglossal nerve is asked to stick out his tongue, the tongue will be pushed toward the right side of the mouth rather than in the midline.

Sensations from the mucous membrane covering the anterior two-thirds of the dorsal surface of the tongue (A) are carried to the brain by the **lingual nerve** (G). Special sensations, such as taste, are carried from the same area by nerve fibers which travel in the lingual nerve but leave the nerve near the base of the skull (H). This is the **chorda tympani nerve,** which later travels with the facial nerve to reach the brain. Ordinary sensations, as well as taste, from the posterior

third of the dorsal surface (B) are carried to the brain by the **glossopharyngeal nerve** (I). The **geniohyoid muscle** (K), although not inserted into the tongue, receives its motor supply from the hypoglossal nerve. The **stylopharyngeus muscle** (L) is closely associated with the glossopharyngeal nerve, from which it receives its motor supply. The middle constrictor muscle of the pharynx (M) will be shown later.

PHARYNX

The various parts of the pharynx are shown in Fig. 93. The **naso-pharynx** (A) is located at

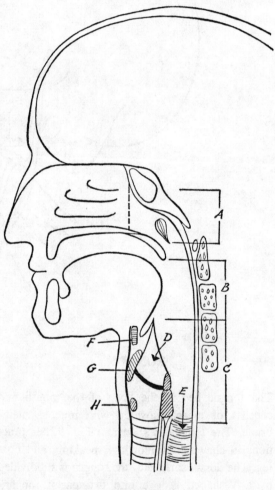

Fig. 93

the posterior part of the nasal cavity and communicates behind the soft palate with the **oral pharynx** (B), which leads into the laryngeal part of the pharynx (C). The arrow D leads from the pharynx into the **larynx,** enclosed by the thyroid (G) and cricoid (H) cartilages. The hyoid bone is at F. The arrow E leads from the pharynx into the beginning of the **esophagus.**

The pharynx is surrounded by a muscular wall that extends from the base of the skull to the cricoid cartilage (Fig. 94). The muscles of the

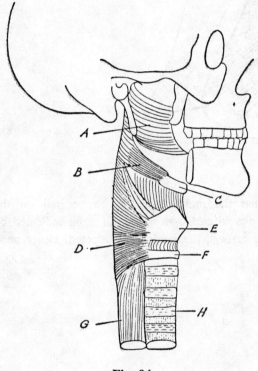

Fig. 94

right and left halves of the tube meet in the midline posteriorly. Although they form a complete muscular wall, it has been customary to separate them into three parts: the **superior constrictor** (A), the **middle constrictor** (B), inserting into the hyoid bone (C), and the **inferior constrictor** (D), inserting into the thyroid (E) and cricoid (F) cartilages. The esophagus (G) and trachea (H) are also shown.

LARYNX

In order to understand the larynx, first look at its skeleton (Fig. 95). The upper structure is

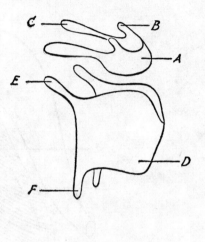

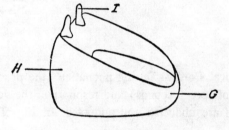

Fig. 95

the body of the hyoid bone (A) with its **lesser** (B) and **greater** (C) **horns.** The middle structure is the thyroid cartilage. Its flat lateral wall, or **lamina** (D), has a **superior** (E) and an **inferior** (F) **horn.** The inferior horn articulates with the cricoid cartilage, which looks like a signet ring. The arch of the cricoid (G) faces anteriorly, and the much larger posterior wall (H) supports two pyramid-shaped cartilages (I) called the **arytenoid cartilages.**

Looking into the larynx from above (Fig. 96), the thyroid cartilage (A) lies superior to the cricoid cartilage (B). Attached to the posterior wall, or lamina, of the cricoid are the **arytenoid cartilages** (C), which have two processes, one pointing anteriorly and one laterally. Sweeping upward and inward from the cricoid cartilage are two fibrous sheets (E) called the **crico-thyroid ligaments.** The upper free edge of each ligament is attached anteriorly to the deep surface of the thyroid cartilage (D) and to the **vocal process** of the arytenoid cartilage. The free edge of this ligament (F) is called the **vocal ligament,** and the mucosa covering the vocal ligaments form the true **vocal cords.**

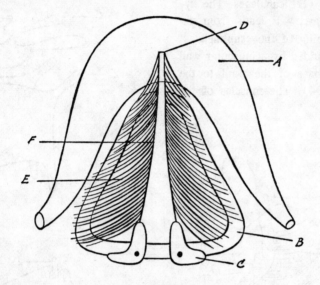

Fig. 96

Vocal Cords. During phonation (the production of speech sounds) and respiration, the vocal cords are abducted and adducted. In Fig. 97A, when the muscles of the larynx pull on the arytenoid cartilages in the direction indicated by the arrows, the cords are adducted. Other mus-

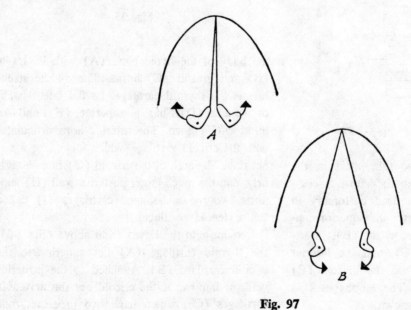

Fig. 97

cles pulling in the opposite direction (Fig. 97B) produce abduction of the vocal cords. To alter the pitch of the voice, a change in the tension of the vocal cords takes place. Since the vocal cords are attached anteriorly to the thyroid cartilage (A in Fig. 98) and posteriorly to the cricoid

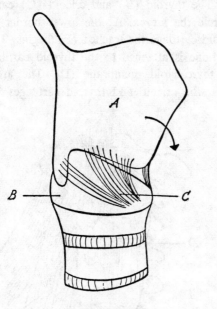

Fig. 98

lamina (B), the crico-thyroid muscle (C), by pulling the thyroid cartilage in the direction of the arrow, will increase tension in the vocal cords.

DEEP STRUCTURES OF THE NECK

Muscles. The deep structures of the neck are shown in Fig. 99. The **atlas** (A) and the **axis** (B) are the first two of the seven cervical vertebrae. Two pairs of muscles (prevertebral) run longitudinally on the bodies of the vertebrae. The **longus colli** (C) and the **longus capitis** (D) flex the neck. The longus capitis muscles (the right one is not shown) and the short **atlanto-occipital muscles** (E) also flex the head, as in nodding "yes." From the transverse processes of the cervical vertebrae the **scalene muscles** (F and G) pass

downward on each side of the neck to be inserted on the ribs. The **scalenus anterior muscle** (G) inserts on the first rib. The **scalenus medius**

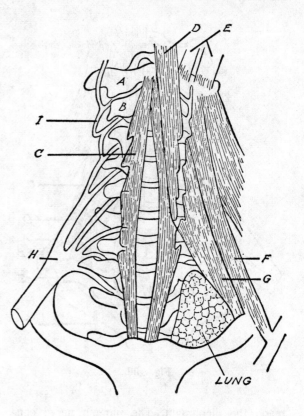

Fig. 99

and **posterior muscles** (F) cannot be separated, and they insert on the first and second ribs. The scalene muscles flex the neck. The prevertebral and scalene muscles receive their nerve supply from the **cervical plexus** of nerves (I). Emerging from the lower half of the vertebral column are the roots of the **brachial plexus** (H).

Blood Supply. On a slightly more anterior plane (Fig. 100) the vessels and nerves in the root of the neck can be seen. The brachial plexus (A) and the adjacent **subclavian artery** (B) are seen emerging from between the scalenus medius (C) and scalenus anterior (D) muscles. The **subclavian vein** (E) is lying anterior to the scalenus anterior muscle, which separates it from the artery. The **phrenic nerve** (F), arising from the cervical part of the spinal cord, lies on the scalenus anterior muscle (D). The nerve passes inferiorly in the neck and into the thorax to

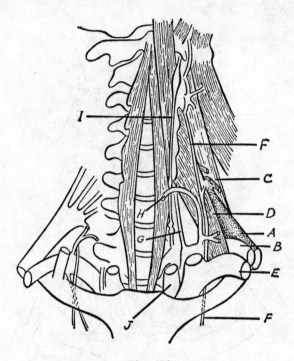

Fig. 100

air reaches the larynx via the pharynx. The hyoid bone (A), a part of the pharynx, provides attachment for muscles coming from the skull as well as from the sternum. These muscles can move the pharynx and larynx up and down, which can readily be observed during swallowing. The thyroid (B) and cricoid (C) cartilages encircle the larynx. At the lower border of the cricoid cartilage the trachea (F) begins. The hyoid bone is attached to the thyroid cartilage by the thyro-hyoid membrane (D). The area between the cricoid and thyroid cartilages (E) is

reach the diaphragm. The phrenic nerve contains the motor fibers that produce contraction of the muscular diaphragm during inspiration. Arising from the subclavian artery in the root of the neck are two important arteries. One is the **vertebral artery** (G), which ascends to the foramen magnum to reach the brain. The other is the **inferior thyroid artery** (H), which sends blood to the thyroid gland. The **sympathetic trunk** (I) lies deep in the neck. Three swellings, or ganglia, are found on the trunk, the largest being the most superior. The other large artery supplying the brain also starts in the root of the neck. It is the common carotid artery (J).

Air and Food Passages. Lying anterior to the prevertebral muscles are the air and food passages (Fig. 101). The air passage consists of the larynx and trachea while the food passage consists of the pharynx and esophagus. However,

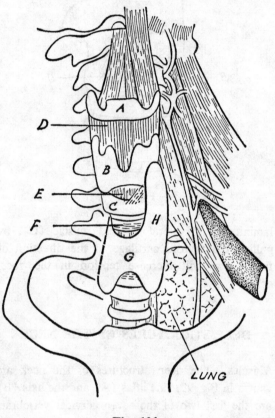

Fig. 101

covered by the crico-thyroid ligament, which is continuous with the crico-vocal membrane described previously.

The **thyroid gland,** the largest endocrine gland found in the body, consists of two lobes. Only the left one (H) is shown. These two lobes are united across the anterior surface of the trachea by the **isthmus** (G). Tracheotomy, an opening into the air passage, is usually done by cutting into the trachea just above or just below the thyroid isthmus.

SUPERFICIAL STRUCTURES OF THE NECK

The anterior surface of the neck is covered by thin **strap muscles** (Fig. 102), which lie super-

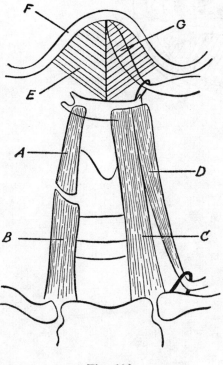

Fig. 102

ficial to the air passage and thyroid gland. They are named according to their attachments, the **thyro-hyoid** (A), the **sterno-thyroid** (B), the **sterno-hyoid** (C), and the **omo-hyoid** (D) muscles. Muscles C and D were removed from the right side to expose the deeper muscles, A and B. The **mylo-hyoid muscle** (E) forms the floor of the mouth and is partially covered by the anterior belly of the **digastric muscle** (G). These

muscles, by pulling on the mandible (F), will depress it, that is, open the mouth. The most superficial muscles of the neck (Fig. 103) are

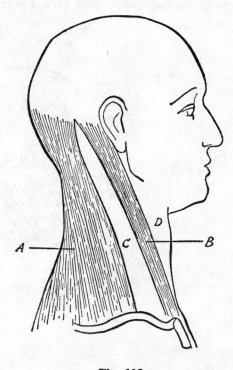

Fig. 103

the **trapezius** (A) and the **sterno-mastoid** (B) muscles. These muscles enclose the posterior triangle of the neck (C). The anterior triangle of the neck (D) is outlined by the sterno-mastoid muscle, the lower border of the mandible, and the midline of the neck.

In a transverse section of the neck (Fig. 104) the relationships of the structures may be reviewed. Lying on the vertebral bodies are the prevertebral muscles (A). Arising from the transverse processes of the cervical vertebrae are the scalenus anterior (B) and the scalenus medius and posterior (C) muscles. Between these two muscle groups the roots of the spinal nerves emerge (J). The trachea (D) is covered on the front and sides by the thyroid gland, and the esophagus lies posterior to the trachea. The common carotid artery, the internal jugular vein, and the vagus nerve, are surrounded by a fibrous sheath (E) that separates these structures from the sympathetic trunk (F). Anterior to the thyroid gland are the strap muscles (G). The most superficial muscles are the trapezius (H) and the sterno-mastoid (I). Superficial to the sterno-mastoid is the external jugular vein (L). Between the trapezius and the vertebrae are the posterior muscles of the neck (K).

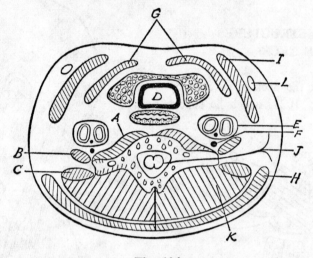

Fig. 104

CHAPTER 7

THORAX, ABDOMEN, AND PELVIS

The trunk contains the three major cavities of the body: the **thoracic cavity,** the **abdominal cavity,** and the **pelvic cavity.** Usually these three regions are described separately, but because of their close interrelationship we will consider them together.

In a sagittal section through the trunk (Fig. 105) the three major cavities are indicated. The thoracic cavity (A) contains the heart, the lungs, and the food and air passages. This cavity is surrounded by the ribs, which form a bony cage. The ribs are attached posteriorly to the **vertebrae** (C) and anteriorly to the **sternum** (B); a muscular septum or wall called the **diaphragm** (D) separates the thoracic cavity from the abdominal cavity (E). Although the ribs form part of the side wall of the abdominal cavity, the major portion of its walls is muscular. The **lumbar vertebrae** (I) form a rigid support in its posterior wall. The abdominal cavity is directly continuous with the cavity of the pelvis (F). A horizontal plane passing from the **pubic symphysis** (G) to the anterior surface of the **first sacral vertebra** (H) is the arbitrary division between the abdominal and pelvic cavities.

Lining Membranes. One feature characteristic of all three cavities is their lining membranes

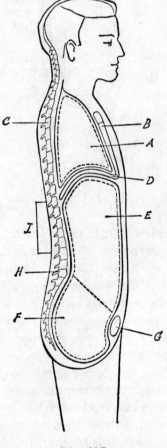

Fig. 105

(Fig. 106). The outer lining of each cavity consists of a dense sheet of connective tissue, or fascia. In the thoracic cavity it is called the **endothoracic fascia** (A), and it covers all walls and the floor of the thoracic cavity, which is

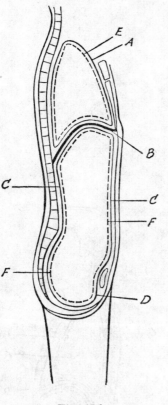

Fig. 106

the diaphragm (B). In the abdominal cavity this fascial layer covers the inferior surface of the diaphragm. Observe that the diaphragm forms the floor of the thoracic cavity as well as the roof of the abdominal cavity. The anterior, lateral, and posterior walls of the abdominal cavity are similarly lined with a fascial layer (C), which then extends inferiorly to line the pelvic cavity (D). This fascial layer, lining both the abdominal and pelvic cavities, is one continuous sheet, often called the **transversalis fascia.** Sometimes the fascial layer takes its name from the particular muscle it is covering. In all three cavities it should be visualized as sticking tightly to

the deep surface of those muscles that form the walls of the cavity.

Adhering closely to the inner surface of the fascial layer is a thin, transparent membrane with a smooth, glistening inner surface. The inner surface of this membrane is thus ideally suited to lessen friction when rubbed by organs within the cavity. In the thoracic cavity this membrane is called **pleura.** Because it lines the walls of the thoracic cavity it is referred to as **parietal pleura** (E) (Latin, PARIES, "wall"). When we wish to indicate a particular area of parietal pleura, we refer to that pleura covering the diaphragm as **diaphragmatic pleura. Costal pleura** is that portion of parietal pleura that lines the side walls of the thoracic cavity.

The thin, smooth layer (F) in the abdominal and pelvic cavities is **peritoneum.** Since it covers the walls of each cavity, it is called **parietal peritoneum.**

Compartments of the Thoracic Cavity. In Fig. 107 the anterior wall of the thoracic and abdominal cavities has been removed. It will be seen that the thoracic cavity is divided into compartments. The **right pleural sac** (A), formed by parietal pleura, contains the right lung. The left lung is contained within the **left pleural sac** (B). The wall or septum separating the two pleural sacs is formed by the remaining organs of the thoracic cavity. The more anterior organs forming the septum are the **heart** (C) with the **thymus gland** (D) just anterior to it. Posterior to the heart are the food and air passages and the great vessel from the heart, the aorta. The cut ends of the ribs (E) indicate that they not only form the bony cage of the thorax but are also part of the lateral walls of the abdominal cavity. The diaphragm (F) forms the partition between the thoracic and abdominal cavities. Note that the convex surface of the diaphragm projects into the thoracic cavity. In the abdominal cavity the anterior surface of the **liver** (G) occupies the upper right area of the abdominal cavity. The **stomach** (H) is found in the upper left area of the same cavity. From the lower border of the stomach an apronlike layer of peritoneum called the **greater omentum** (I) covers most of the intestines. From this angle we can look into the pelvic cavity (J), although no pelvic organs are shown in this diagram.

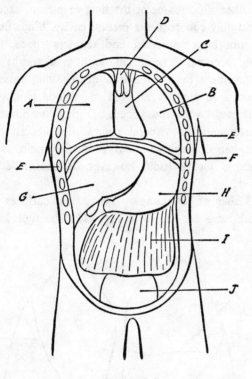

Fig. 107

In a transverse section of the thorax (Fig. 108) the compartments of the thoracic cavity are shown. The bony cage is formed by the sternum (A) in front, by the ribs (B) on the sides, and posteriorly by the vertebrae (C). The right (D) and left (E) pleural sacs are formed by the parietal pleura and are separated by a septum, (F), or wall, formed by the heart (G), the air passage or trachea (H), the food passage or esophagus (I), and the great vessel from the heart, the aorta (J). The septum in the thorax is called the **mediastinum.** The space anterior to the heart (K) is called the **anterior mediastinum;** it contains some fat and a few lymph nodes. The **middle mediastinum** is that space filled by the heart

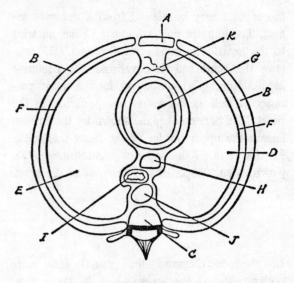

Fig. 108

and its coverings. The **posterior mediastinum** contains those structures lying posterior to the heart. According to the general scheme for naming areas of parietal pleura, observe that the layer of parietal pleura covering the side walls of the mediastinum is called **mediastinal pleura.**

In a sagittal section through the middle of the mediastinum (Fig. 109) the components of

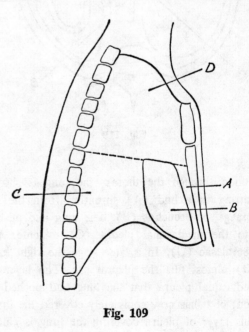

Fig. 109

the septum may be viewed from a different aspect. The anterior mediastinum (A) lies anterior to the heart. The middle mediastinum (B) contains the heart and its coverings. The posterior mediastinum (C), containing food and air passages as well as the aorta, lies posterior to the heart. The horizontal plane shown by the broken line at the upper border of the heart marks the inferior limit of the superior mediastinum (D), which extends superiorly to the root of the neck.

LUNGS

In previous diagrams the pleural sacs were shown, although the lungs, which fill those sacs, were not indicated. In Fig. 110, a horizontal

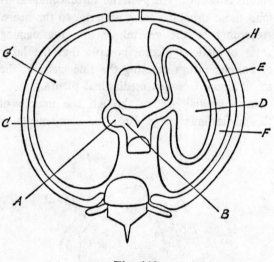

Fig. 110

visceral pleura (E). The space between the visceral and parietal pleura (H) is called the **pleural cavity** (F). Normally the space is very small since the visceral and parietal layers are barely separated by a film of fluid. This fluid helps to reduce friction as the two layers rub against one another. In diseases of the lung or pleura, excess fluid may collect in the pleural cavity. Since fluid cannot be compressed and requires space, the lung becomes compressed, and, as a result, an impairment in the function of the lung usually occurs. In pleurisy an area of parietal pleura is inflamed and roughened. During respiratory movements the visceral pleura rubs against the roughened sensitive area, producing pain. The visceral pleura itself, however, is insensitive to pain.

Lobes of the Lungs. The lateral surfaces of both lungs are shown in Fig. 111. The right lung

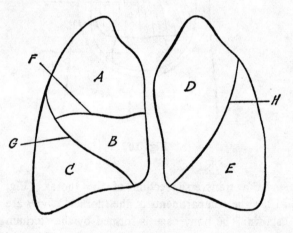

Fig. 111

section through the thorax, the lung is shown starting as a bud (A) sprouting from the air passage, or **bronchus** (B). The lung bud pushes into the mediastinal pleura (C) to enter the pleural sac (G). In a later stage the right lung (D) almost fills the pleural sac. The layer of mediastinal pleura that the lung had pushed in front of it has now completely covered the lung. This layer of pleura covering the lung is called

is divided into three lobes. The **upper lobe** (A), **middle lobe** (B), and **lower lobe** (C) are separated by two deep fissures, the **horizontal fissure** (F) and the **oblique fissure** (G). The left lung consists of an upper lobe (D) and a lower lobe (E), separated by the oblique fissure (H). The medial surfaces of the lungs are shown in Fig. 112. The lobes of each lung are indicated as before. The trachea (F) divides into the right (G) and left (H) **bronchi** (singular, bronchus).

In the area where the bronchi enter the lungs the cut ends of two other structures are seen. The **pulmonary vein** (I) carries oxygenated blood back to the heart, while the **pulmonary artery** (J) carries blood to the lungs to be oxygenated. The oval area where all three structures either enter or leave the lung is called the **lung root.**

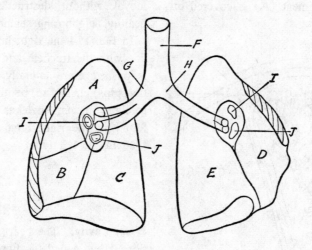

Fig. 112

Lung Segments. Each lung is further subdivided into distinct anatomical units called **segments.** The upper lobe of the right lung (Fig. 113) has three segments: X, Y, and Z. In Fig. 114 the right bronchus (F) divides into smaller bronchi; one goes to each of the three segments in the upper lobe (A). The middle lobe of the right lung (B) contains two segments, and the

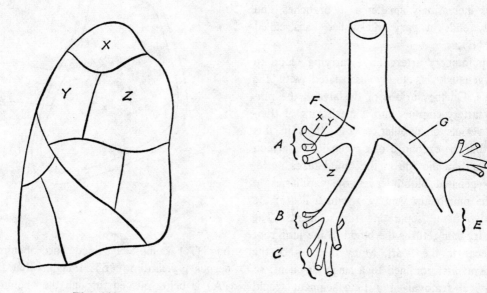

Fig. 113

Fig. 114

lower lobe (C) has five segments. The left bronchus (G) also divides to supply the segments of
the upper (D) and lower (E) lobes.

A schematic diagram of a segment is shown
in Fig. 115. A lung segment (A) is covered on

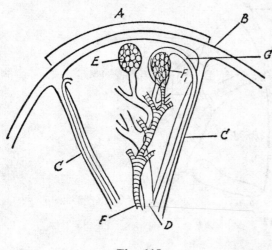

Fig. 115

its periphery by visceral pleura (B), which also
covers the periphery of adjacent segments. Segment A is separated from adjacent segments by
fibrous connective tissue (C). The bronchus (D)
becomes increasingly smaller as it branches, and
it finally ends in very thin-walled sacs called
alveoli (E).

The pulmonary artery (F), carrying blood to
be oxygenated, is closely associated with the
bronchus. On the surface of the alveoli the pulmonary artery empties into a meshwork of thin-
walled vessels or capillaries (F_1). It is in this
region that the exchange of oxygen in the alveoli
for the carbon dioxide in the blood takes place.
The oxygenated blood leaves the capillaries to
enter the pulmonary vein (G), which lies in the
connective tissue separating the segments. The
pulmonary vein returns the blood to the lung root
and thence to the heart. Many diseases of the
lung are at first confined to a single segment, so
the surgical removal of a lung segment would
be preferable to the removal of a lobe. Since

each segment has its own air and blood supply,
these structures may be tied off without affecting
adjacent segments. The fibrous connective tissue
surrounding a segment permits its surgical removal without destruction of lung tissue in
healthy neighboring segments.

In Fig. 114 the right bronchus (F) is more in
line with the trachea, and the left bronchus (G)
is pushed more laterally. This anatomical relationship appears to be the reason why inhaled
foreign objects, such as buttons or peas, are
found in the right lung much more frequently
than in the left lung.

THE HEART

Heart Cavity. The heart is contained in a tough
fibrous bag called the **fibrous pericardium,** which
is lined with a thin, glistening membrane called
serous pericardium. Serous pericardium is similar to pleura and peritoneum in structure as well
as in function. In order to visualize the relationship of the pericardium to the heart, first observe
the fibrous pericardium (Fig. 116) as an open

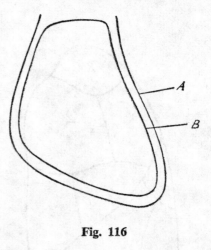

Fig. 116

bag (A) completely lined (no opening) with
serous pericardium (B). Imagine that the heart
(A) is being pushed through the opening of the
bag (Fig. 117) and is pushing in front of it a

layer of serous pericardium (B). The fibrous bag (C) is lined with serous pericardium (D). The space (E) will be the **pericardial cavity.**

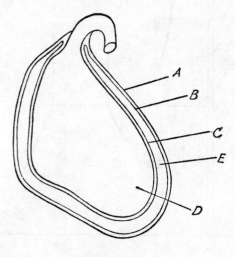

Fig. 117

The final relationship is shown in Fig. 118. The fibrous pericardium (A) is lined with serous pericardium (B). A continuation of this serous layer (C) adheres closely to the heart muscle (D). The space between the two adjacent layers of serous pericardium is the pericardial cavity (E). Since the fibrous bag is tough and unyielding, excess fluid or blood in the pericardial cavity

Fig. 118

may seriously interfere with the pumping action of the heart muscle.

Parts of the Heart. The adult heart weighs approximately three hundred grams in the male and 250 grams in the female. (There are 454 grams to the pound.) A front view of the heart is shown in Fig. 119. The **right atrium** (A) and

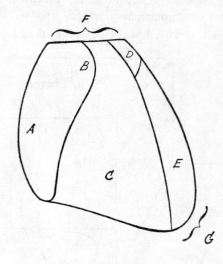

Fig. 119

its earlike process (B), the **auricle,** form the right border of the heart. The **right ventricle** (C) takes up most of the anterior surface, while the **left ventricle** (E) forms most of the left border. The auricle of the **left atrium** (D) is just visible at the upper left corner since the left atrium is best seen from the posterior aspect. The base of the heart is indicated at F, and the apex of the heart is indicated at G.

Flow of Blood through the Heart. Since it is difficult to diagram the flow of blood through the chambers of the heart in their actual relationships, a schematic diagram of the heart is shown in Fig. 120. The heart consists of four chambers. The right and left atria are separated by the **interatrial septum;** the right and left ventricles are separated by the **interventricular septum.** Unoxygenated (venous) blood from the head, neck, and thoracic wall is carried to the right atrium by the **superior vena cava** (A). Venous blood from the rest of the body empties into the right atrium through the **inferior vena cava (B)**. Contraction of the right atrium forces the blood through the **tricuspid valve (C)** into the right ventricle. Contraction of the right ventricle forces the blood through the **pulmonary valve (D)** into the pulmonary trunk and thence through the pulmonary arteries to the lungs for aeration. The tricuspid and pulmonary valves are so constructed that in a healthy heart blood can pass only in one direction. The oxygenated blood is returned from the lungs to the left atrium by the four pulmonary veins (E), two from each lung. Contraction of the left atrium forces the blood through the **mitral valve (F)**, a two-cusp valve, into the left ventricle. Contraction of the left ventricle forces the blood through the **aortic valve (G)** into the aorta, which carries the oxygenated blood to all parts of the body. The mitral and aortic valves also allow blood to flow in only

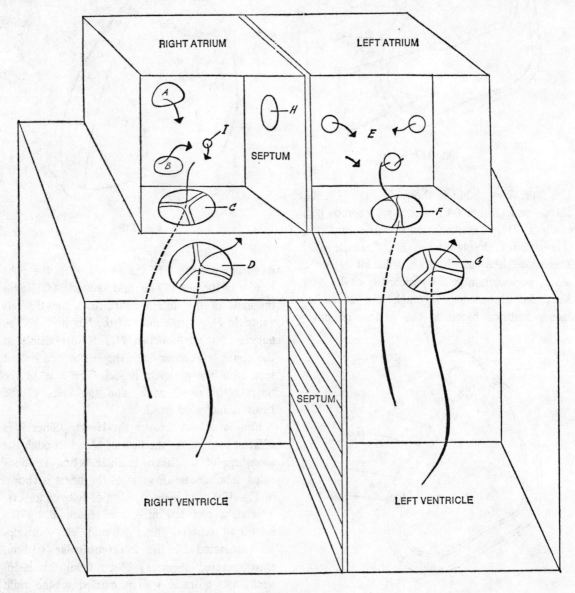

Fig. 120

one direction. The cycle is completed by the superior and inferior venae cavae returning the unoxygenated blood to the right atrium.

In the septum between the atria an oval depression called the **fossa ovale** (H) marks the site of an actual opening found in the fetal heart, the **foramen ovale.** The opening permits blood to pass directly from the right atrium into the left atrium. While the foramen ovale usually closes after birth, its persistence is one type of congenital heart defect. The opening (I) in the right atrium is the end of the **coronary sinus,** a large vein that drains venous blood from the major part of the heart muscle.

Blood Vessels That Work with the Heart. The major blood vessels associated with the heart are shown in Fig. 121. The superior vena cava

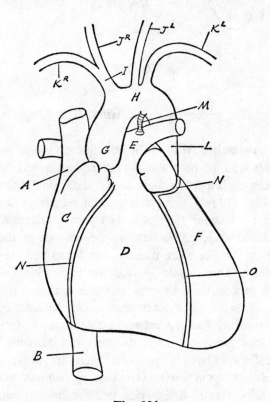

Fig. 121

(A) and the inferior vena cava (B), carrying unoxygenated blood, empty into the right atrium (C). Blood from the right ventricle (D) is

pumped into the pulmonary trunk (E), which divides into right and left pulmonary arteries to the corresponding lung. The left ventricle (F) pumps oxygenated blood into the ascending aorta (G). From the arch of the aorta (H) three large arteries arise to supply the upper limbs, head, and neck. The **brachiocephalic artery** (I) divides into the **right subclavian artery** (K^R) going to the upper limb, and the **common carotid artery** (J^R) going to the right side of the head and neck. The left common carotid artery (J^L) and the left subclavian artery (K^L) arise directly from the arch of the aorta and supply corresponding left-sided structures. The descending aorta (L) continues inferiorly behind the heart through the thoracic cavity into the abdominal cavity. A cordlike structure, the **ligamentum arteriosum** (M), is seen connecting the arch of the aorta with the pulmonary trunk. In the fetus an actual communication, or shunt, called the **ductus arteriosus,** exists between the aorta and the pulmonary trunk. Through this communication, blood from the right ventricle is shunted into the aorta since the fetal lungs do not function. After birth, the ductus arteriosus usually closes off, becoming a cordlike structure. Failure to close results in a patent, or open, ductus arteriosus, which is another type of congenital heart defect. The groove (N) indicates the line of separation between the atria and ventricles and is called the **atrio-ventricular sulcus.** The main trunks of the coronary arteries lie in this sulcus. The groove (O) on the anterior surface of the heart marks the site of the interventricular septum and is occupied by a branch of the left coronary artery.

The coronary arteries (Fig. 122) arise from the aorta just above the aortic valve (A). The **right coronary artery** (B) runs in the atrio-ventricular sulcus and turns around the right border of the heart to lie on its posterior aspect (C), where it sends the posterior interventricular branch (D) inferiorly to supply the posterior surfaces of both ventricles. The **left coronary artery** (E) soon divides into a circumflex branch (F), which turns around the left border of the heart to supply the posterior surface. The other branch (G), the anterior interventricular artery, descends to supply the anterior surfaces of both ventricles.

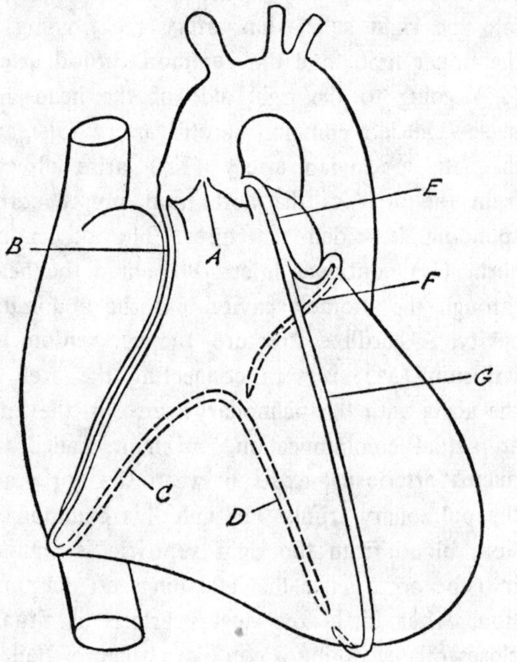

![Fig. 122]

Fig. 122

ABDOMEN

Diaphragm. The diaphragm separating the thoracic from the abdominal cavity is part muscular and part tendinous. In Fig. 123 the diaphragm is illustrated as seen from its abdominal or concave surface. The muscular part (C) arises from the **costal arch** (A) as well as from the xiphoid process of the sternum (B). The muscular fibers are inserted into the **central tendon** (D). The inferior vena cava, after passing superiorly in the abdominal cavity, pierces the tendinous part of the diaphragm (E). The esophagus descends in the thoracic cavity to pierce

the diaphragm (G) just before it joins the stomach. The descending aorta (F) passes under the arch formed by the diaphragm and the vertebral bodies. The diaphragm is anchored firmly to the lumbar vertebra (I) by two strong, fibrous,

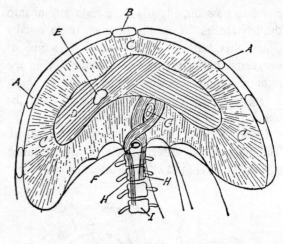

Fig. 123

straplike processes called **crura** (H) (singular, crus).

Abdominal Wall. The muscles of the anterior wall of both thorax and abdomen will be presented together because of their similarities (Fig. 124). The most superficial muscle in the spaces between the ribs is the **external intercostal muscle** (A), shown only on the left side of the body. Its fibers are directed downward and medially. This muscle is vital for respiration because it elevates the ribs and thus increases the capacity of the thoracic cavity. On the anterior abdominal wall the **external oblique muscle** (B) is the most superficial; its fibers run downward and medially. The muscular fibers are inserted into its **aponeurosis** (C). This aponeurosis extends to the midline, where its fibers interdigitate with the right external oblique aponeurosis, which is not shown here. The inferior border of the external oblique aponeurosis is folded under, forming a cordlike structure called the **inguinal ligament** (D). The opening in the aponeurosis

(E) is the **external inguinal ring**. It is through this opening that the spermatic cord passes on its way from the testes to the pelvic cavity. A pair of straplike muscles (J) are outlined since they lie deep to the aponeurosis. These muscles arise from the pubic symphysis (F) and are attached to the sternum.

On the right side of the body the first muscle

layer has been removed to reveal the second layer. The fibers of the **internal intercostal muscle** (G) run upward and medially. The muscle fibers of the corresponding layer of the abdominal wall, the **internal oblique muscle** (H), run in the same direction. The aponeurosis of the internal oblique muscle (I) covers the **rectus abdominis muscle**. The third and deepest muscle

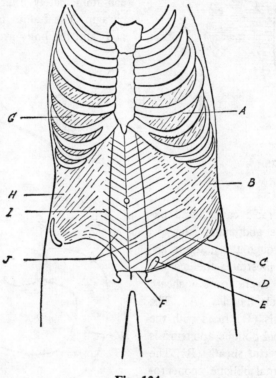

Fig. 124

layer is shown on the left side of the body in Fig. 125. The innermost intercostal muscle fibers (A) run obliquely, whereas the **transversus abdominis muscle** fibers (B) run transversely, as the name implies. However, both the intercostal (E) and abdominal wall (F) nerves lie just superficial to this layer. The greater part of the aponeurosis of the transversus abdominis (C) passes deep to the rectus abdominis muscle (G), thus exposing it. However, the most inferior part

of the aponeurosis (D) passes superficial to the rectus.

On the right side of the body, after removal of the third muscle layer, the fascial layer is revealed. In the thorax it is called the **endothoracic fascia** (I), and in the abdominal wall it is known as the **transversalis fascia** (H). The opening (J) in the transversalis fascia indicates the site of the internal inguinal ring through which the spermatic cord passes. Deep to the fascial layer pa-

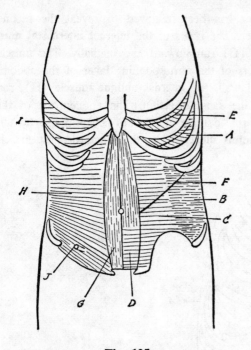

Fig. 125

(G) to form the **posterior rectus sheath (C).** The aponeurotic fibers interdigitate with their opposite numbers in the midline (H) to form the **linea alba,** or white line. I is the transversalis fascia and J represents the parietal peritoneum.

STOMACH AND DUODENUM

In Fig. 127 the esophagus (A) joins the stomach immediately after passing through the diaphragm. The **fundus** of the stomach is indicated at B and the **body** at C. The pyloric part of the

rietal pleura lines the thoracic cavity, and parietal peritoneum lines the abdominal cavity.

A transverse section through the anterior abdominal wall (Fig. 126) illustrates the manner in which the aponeurotic layers form a sheath around the rectus abdominis muscles (A). The external oblique aponeurosis (D) fuses with the anterior sheet of the internal oblique aponeurosis (E) to form the **anterior rectus sheath** (B). The posterior sheet of the internal oblique aponeurosis (F) fuses with the transversus aponeurosis

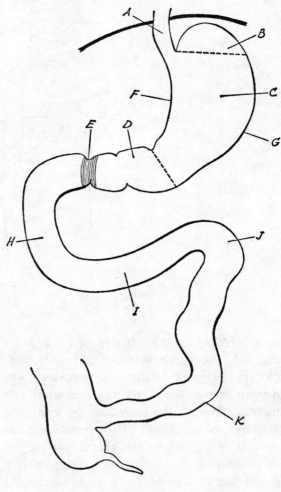

Fig. 127

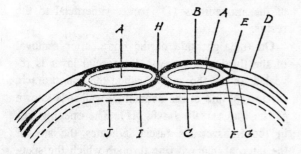

Fig. 126

stomach (D) narrows as it approaches the **pyloric sphincter** (E), a muscular collar which acts like a purse string when contracted, to prevent emptying of the stomach contents into the duodenum. The **lesser** and **greater** curvatures of the stomach are indicated at F and G respectively. The duodenum, which is shaped like the letter *C*, begins just beyond the pyloric sphincter. The first part runs to the right and descends to form the second part (H), which then turns toward the left to form the third part (I). At J the duodenum makes a sharp turn, and this part of the small intestine is known as the **jejunum.** The terminal part of the small intestine is called the **ileum** (K). The lining of the stomach (Fig. 128) is thrown into many folds, called **rugae** (A), which increase its surface area. The pyloric sphincter (B) is formed by a ring of circular muscle fibers that marks the junction between the stomach and the duodenum.

Pancreas. In Fig. 129 the head of the pancreas (A) fits into the concavity of the duodenum. The body (B) and tail (C) of the pancreas extend toward the left side of the abdominal cavity. By removing some of the glandular tissue, we expose the **main pancreatic duct (D).** The secretions of the glandular tissue are transported by the duct toward the head of the pancreas. Just before the duct enters the duodenum

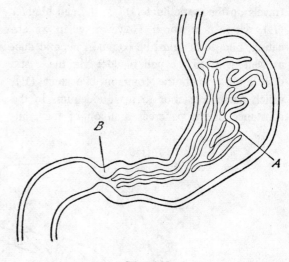

Fig. 128

it unites with the **common bile duct** (E) to form an enlarged channel, or **ampulla** (F). The relationship of the pancreatic duct to the common

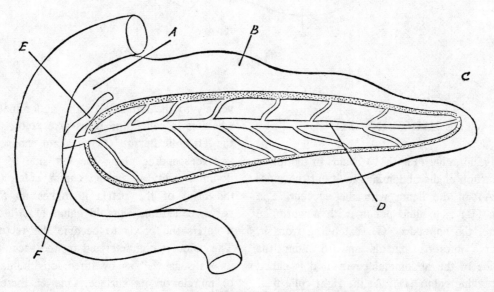

Fig. 129

bile duct may be visualized better in Fig. 130. Bile is produced in the liver and is drained from the liver by the **left hepatic duct** (A) and **right hepatic duct** (B), which unite to form the **common hepatic duct** (C). Usually the bile then travels up the **cystic duct** (D) to the **gall bladder** (E), where it is stored. However, when we are eating, bile flows from the common hepatic duct as well as from the gall bladder via the cystic duct and through the common bile duct (F), which passes posterior to the duodenum. In the substance of the pancreas a union of the common bile duct with the pancreatic duct (G) takes place. This common channel, the ampulla (H), pierces the wall of the duodenum and discharges both bile and pancreatic secretions into the lumen of the **gut** (I). Blockage of bile flow in the ducts may occur in several areas. A stone from the gall bladder may be caught in the common bile duct. Cancer of the head of the pancreas may block the common bile duct as it passes through the substance of the gland. If the bile flow is blocked, the level of the bile pigments in the blood increases, producing jaundice.

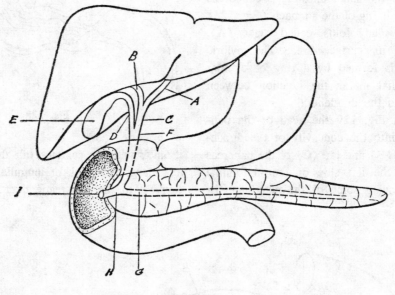

Fig. 130

THE LARGE INTESTINE

The large intestine (Fig. 131) begins in the lower right portion of the abdominal cavity at the junction (A) of the ileum with the **caecum.** The caecum (B) is a blind pouch, with a wormlike structure, the **appendix** (C), extending from it. The large intestine extends upward along the right side of the abdominal cavity and is called the **ascending colon** (D). At the **right colic flexure** (E) it changes direction and passes transversely to the left side of the abdominal cavity, where it is called the **transverse colon** (F). At the **left colic flexure** (G) it again changes direction, descending along the left side of the abdominal cavity as the **descending colon** (H). At the brim of the pelvis it makes an S-shaped curve, called the **sigmoid colon** (I), just before it enters the pelvis to become the **rectum** (J). The colon is characterized by the sacculation of its walls as well as by three longitudinal bands of muscle on its surface. One of these bands, the **taenia coli** (K), is shown.

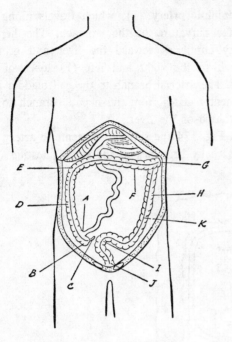

Fig. 131

BLOOD SUPPLY TO THE ABDOMEN

Abdominal Arteries. The abdominal portion of the aorta (Fig. 132) emerges from under the arch of the diaphragm (A) and very shortly thereafter gives rise to the **coeliac axis,** or coeliac trunk (B). The trunk is usually very short; it

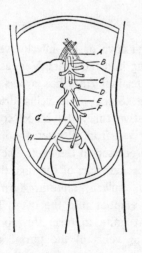

Fig. 132

divides into three branches, which supply the stomach, spleen, liver, and part of the duodenum and pancreas. Slightly inferior to the coeliac axis, the **superior mesenteric artery** (C) arises from the aorta. It supplies the rest of the small bowel as well as approximately the first half of the large bowel. At a slightly lower level of the aorta the **renal arteries** (D) arise. Since the testes originate near the kidney region in the embryo, their arterial supply, the **testicular arteries** (E), arises from the aorta near the renal arteries. The remaining large artery arising from the aorta is the **inferior mesenteric** (F), which supplies the remaining part of the large intestine. At the level of the fourth lumbar vertebra the aorta divides, or bifurcates (G), forming the **right** and **left common iliac arteries.** The common iliac artery divides (H) to form the **external iliac artery,** which supplies the lower limb, and the **internal iliac artery,** which runs into the pelvis to supply the structures found there.

A general plan of the arterial branches of the coeliac axis is shown in Fig. 133. The coeliac

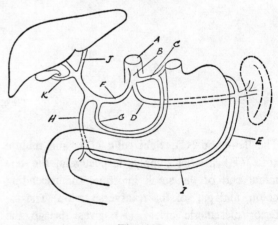

Fig. 133

trunk (B), arising from the aorta (A), divides into the **left gastric artery** (C), which runs along the lesser curvature of the stomach. Another branch of the coeliac trunk is the **splenic artery** (D). A branch of the splenic artery, the **left gastro-epiploic artery** (E), runs along the greater curvature of the stomach. A third branch of the coeliac trunk is the **hepatic artery** (F), which gives rise to the **right gastric artery** (G). Both right and left gastric arteries anastomose with each other on the lesser curvature of the stomach. Another branch of the hepatic artery is the **gastro-duodenal artery** (H), which sends branches to the duodenum and pancreas. The gastro-duodenal artery gives rise to the **right gas-** **tro-epiploic artery** (I), which travels along the greater curvature of the stomach. The hepatic artery continues toward the liver and gives a branch to the right and left (J) lobes of the liver. The arterial branch to the gall bladder (K) frequently arises from the hepatic branch to the right lobe of the liver.

In Fig. 134 the superior mesenteric artery (A) gives many branches to the small intestine (B).

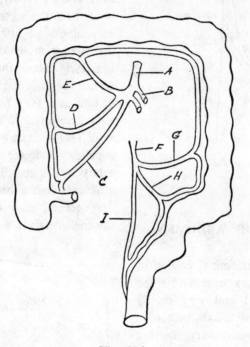

Fig. 134

The **ileo-colic** (C), **right colic** (D), and **middle colic** (E) arteries supply, respectively, the terminal end of the small intestine, the ascending colon, and part of the transverse colon. The inferior mesenteric artery (F) gives rise to the **left colic** (G), **sigmoid** (H), and **superior hemorrhoidal** (I) arteries, which supply part of the transverse colon, the descending colon, the sigmoid colon, and the rectum.

Abdominal Veins. Venous blood, that is, deoxygenated blood, from most parts of the body, with the exception of the major part of the gastrointestinal tract, travels eventually to either the superior vena cava or the inferior vena cava before emptying into the heart. Venous blood from the gastrointestinal tract has a less direct route to the heart. Since these veins drain the wall of the digestive tube, the blood contains those products derived from digested food, which must be further processed in the liver. The veins draining the abdominal part of the digestive tract, the pancreas and spleen, unite to form the **portal vein,** which empties into the liver. The network of veins that unite to form the portal vein are referred to as **veins of the portal system** (Fig. 135). Those veins going directly into the superior vena cava and inferior vena cava are part of the **caval system** of veins. The **superior mesenteric vein** (A), which drains the small intestine and part of the large intestine, unites with the **splenic vein** (C) to form the major part of the portal venous system. The **inferior mesenteric vein** (B) from the large intestine empties into the splenic vein. The **gastric vein** (D) drains part of the stomach, as well as the lowest part of the esophagus, and empties into the splenic vein. The superior mesenteric and splenic veins unite to form the portal vein (E), which enters the liver. The blood of the portal vein now flows through a meshwork of thin-walled blood channels called **sinusoids** (F). From the sinusoids the

blood is picked up by the **hepatic veins** (G), which enter the inferior vena cava (H) near its termination in the right atrium.

There are several areas where veins of the portal system anastomose with veins of the caval system. Two areas are shown here, one at the lower end of the esophagus (I) and the other at the lower end of the rectum (J). If the blood pressure within the veins of the portal system is raised, the increased pressure may be transmitted to the veins of the caval system at the sites of anastomoses (I and J). The veins in these two areas are surrounded by loose connective tissue, which does not give much support to the vein wall. Thus, when the pressure within the vein

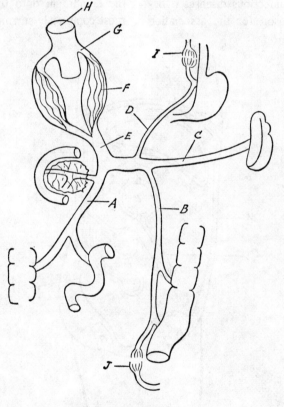

Fig. 135

increases, the vein wall dilates and becomes tortuous, forming a varicose vein. What produces an increased blood pressure in the portal system?

A blood clot, or thrombus, in the portal vein would result in increased pressure in veins A, B, C, and D. If the liver is exposed to toxic substances, the liver cells become swollen; since they lie adjacent to the sinusoids, they may block the flow of blood through the sinusoids, resulting in increased pressure in the portal system.

Spleen. The spleen is normally a soft, friable organ enclosed by an external coat made up of fibrous, elastic, and muscular tissue. It lies inferior to the left leaf of the diaphragm and postero-lateral to the stomach. It contains a system of scavenger cells that ingest almost any type of particle. Worn-out red blood cells are included in this category, and the spleen can temporarily store a considerable volume of blood which can be poured into the circulation when required.

The spleen is the largest lymphoid organ in the body and in certain infectious diseases it becomes enlarged. The reason for the association of the spleen with the portal system is not clear, although some of the raw material required for the manufacture of bile in the liver is derived from the hemoglobin that comes from red blood cells that have broken down.

PERITONEUM

The abdominal cavity shown in Fig. 136 is lined with a very thin, smooth, shiny layer of tissue called peritoneum. The part of the peritoneum that lines the wall is called parietal peritoneum (N). Follow the parietal peritoneal layer upward, where it covers the abdominal surface of the diaphragm (I). This layer is reflected off the diaphragm onto the liver (A), which it almost completely surrounds. This part of the peri-

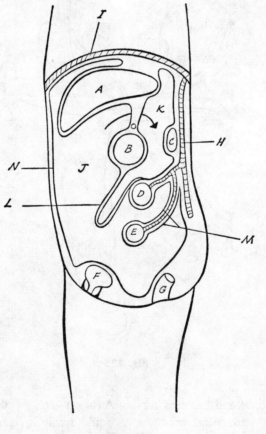

Fig. 136

toneal layer is called **visceral peritoneum** because it covers a **viscus,** or organ. This same layer is reflected off the liver onto the stomach (B) and forms part of a fold of peritoneum hanging from the stomach called the greater omentum (L). Return to the parietal peritoneal layer lining the anterior abdominal wall and follow it downward into the pelvis, where it is reflected onto the bladder (F) and then onto the rectum (G), where it lines the posterior abdominal wall. In a manner similar to that described for the visceral pleura covering the lung, the small intestine (E) is covered by a layer of peritoneum. The double layer of peritoneum (M) suspending the small intestine to the posterior abdominal wall is called the **mesentery.** Between the two layers, blood vessels from the aorta (H) reach the wall of the intestine. Veins, nerves, and lymphatic vessels also run between these two layers of peritoneum. The transverse colon (D) is similarly covered with peritoneum and has its own mesentery, conveying blood vessels, nerves, and lymphatics. The pancreas (C) is covered by the parietal peritoneal layer, but it has no mesentery. Follow this layer upward to the point at which it is reflected off the diaphragm onto the liver, then onto the posterior surface of the stomach. This layer then continues inferiorly to complete the remaining layer of the greater omentum (L).

The spaces J and K are called the **peritoneal cavity.** You will observe that no bare wall of any organ or blood vessel is exposed to the peritoneal cavity. The peritoneal cavity is very small because the organs covered with visceral peritoneum fill the abdominal cavity and are everywhere in contact with the parietal peritoneum. A small amount of fluid in the peritoneal cavity allows easy movement of the viscera. Under some conditions excess fluid, called **ascites,** is produced in the peritoneal cavity. The long mesentery of the small intestine may become twisted, shutting off the blood supply to the wall of the intestine. This condition is called **volvulus.**

The double peritoneal layer suspending the stomach from the liver is called the **lesser omen-tum.** The ability of the lymphatics in the mesentery to carry cancer cells from the stomach to the liver explains secondary cancer growths, or **metastases,** in the liver from the original cancer in the stomach. Infection in the peritoneal cavity, or **peritonitis,** is very serious, perhaps because of the rapidity with which the toxic products of bacteria are absorbed through the peritoneum into the bloodstream. Since rapidity of absorption through a membrane is related to the surface area of that membrane, it is significant that the surface area of the peritoneal cavity is approximately equal to the individual's skin surface area.

The peritoneal cavity is divided into two interconnecting areas by the stomach and its mesentery. The **greater sac (J)** communicates with the **lesser sac (K)** by means of a peritoneal lined channel called the **epiploic foramen,** indicated by the arrow. Thus a perforation of the posterior wall of the stomach would permit stomach contents to enter the lesser sac. Infection, however, can spread to the greater sac through the epiploic foramen.

PELVIS

The relationship of the kidneys to the posterior wall of the abdominal cavity is shown in Fig. 137. The diaphragm (A) forms the upper part of the posterior abdominal wall. It then arches to form the roof of the abdominal cavity. The major part of the posterior abdominal wall is formed by two muscles. The more lateral one is the **quadratus lumborum** (B), which runs from the twelfth rib to the iliac crest. The more medial one is the **psoas major** (C), which unites with the **iliacus muscle** (D) to form the **iliopsoas** muscle.

The origin of the transversus abdominus muscle (E) can be seen as it sweeps around to form the lateral and anterior walls of the abdominal cavity. The aorta (F) is emerging under an arch of the diaphragm where it travels distally on the vertebrae. At the level of the fourth lumbar vertebra it bifurcates, forming the right and left common iliac arteries. The common iliacs again divide, forming the external iliac artery, which skirts the brim of the pelvis to enter the lower limb, and the internal iliac artery, which enters the true pelvis.

The blood supply of the kidney is shown in Fig. 138. The lower pole of the right kidney is at A. The upper pole is capped by the suprarenal gland (B). Arising from the aorta (C), the renal artery enters the kidney, while the renal vein (G) returns the blood to the inferior vena cava (D). Emerging from the kidney is the renal pelvis (E), which is the dilated upper end of the ureter (F), which drains the urine to the bladder.

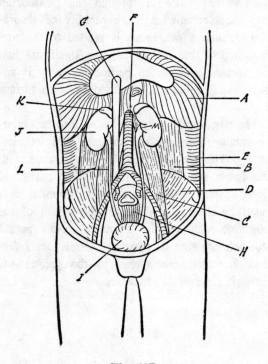

Fig. 137

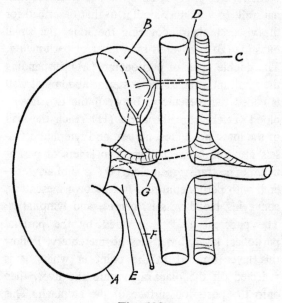

Fig. 138

Two organs are shown within the true pelvis. Near the anterior wall is the **urinary bladder** (I), and near the posterior wall is the stump of the **rectum** (H). The inferior vena cava (G) lies to the right of the aorta and pierces the diaphragm to enter the right atrium of the heart. Its iliac branches (not shown) correspond to the artery of the same name. The kidneys (J) rest against the diaphragm, the psoas major, and quadratus lumborum muscles.

Kidneys. The upper end of each kidney is capped by the **suprarenal gland** (K). The indented area of the kidney (M) is called the **hilum.** Emerging from the hilum is the ureter (L).

If the kidney were bisected longitudinally, the cut surface would resemble Fig. 139. The periphery of the organ is called the **cortex** (A) and contains the filtering system. The **renal pyramids** (B) are made up of small tubules that empty the urine into a **minor calyx** (C) (plural, calyces). Several calyces join to form a **major calyx** (D). Usually, two major calyces unite to form the **renal pelvis** (E), which then narrows to form the ureter (F), a muscular tube which travels distally on the psoas major muscle and crosses the iliac vessels to enter the true pelvis. The ureter ends by piercing the muscular wall of the bladder.

Kidney stones are sometimes formed in the renal pelvis. If the stone is small enough, it may be pushed along the ureter by its muscular wall, resulting in severe pain. Stones may be held up at three constrictions of the ureter: its junction with the renal pelvis, the point where it bends over the brim of the true pelvis, and the point where it enters the muscular wall of the bladder.

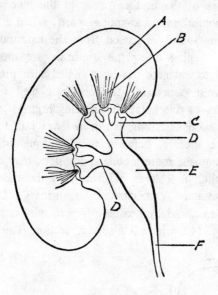

Fig. 139

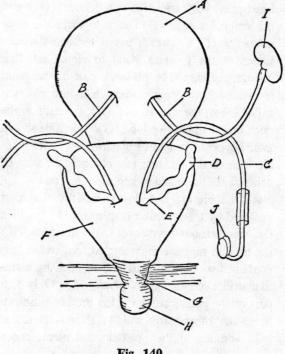

Fig. 140

In a sagittal view of the same structures (Fig. 141) the urinary bladder is shown (A). The prostate gland (B) forms a base for the bladder. The constriction of the bladder cavity (N) marks

Urinary Bladder. The posterior aspect of the bladder and its related structures are shown in Fig. 140. The urinary bladder (A) receives the ureters (B) through its posterior wall. Surrounding the inferior part or neck of the bladder is the **prostate gland** (F). Entering the posterior surface of the prostate are the **vas deferens** (C) and the **seminal vesicles** (D), which unite to form the **ejaculatory duct** (E), which empties into the urethra. The vasa deferens carry sperm from the testes, (J), while the seminal vesicles contribute seminal fluid but apparently do not store sperm.

Urethra. Emerging from the apex of the prostate is the membranous part of the **urethra** (G) surrounded by a muscular diaphragm. Inferior to this muscular diaphragm, the urethra makes a ninety-degree turn anteriorly. The area of the right-angled turn is called the **bulb** of the urethra (H).

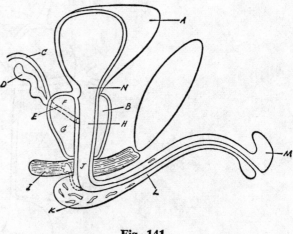

Fig. 141

the beginning of the urethra (H). Since this part of the urethra is surrounded by the prostate, it is referred to as the **prostatic urethra.** Two structures previously seen, the vas deferens (C) and the seminal vesicle (D), unite to form the ejaculatory duct (E), which passes through the substance of the prostate gland to empty into the prostatic urethra. The posterior part of the gland is divided by the ejaculatory duct into two so-called lobes, the **middle lobe** (F), superior to the duct, and the **posterior lobe** (G). It will be shown later how enlargement or hypertrophy of the middle lobe of the prostate can obstruct the flow of urine from the bladder into the urethra.

As the urethra is followed distally, that part enclosed by the muscular diaphragm (I) is called the **membranous urethra** (J). At the bulb (K) the urethra makes a right-angled turn where it is contained within the **penis** (L) to open by a slit-like orifice on the **glans penis** (M). Only that part of the penis containing the urethra is shown here; the other parts will be shown later.

In addition to the prostatic and membranous parts of the urethra, the remaining mobile part is called the **spongy,** or **penile,** urethra. The spongy part of the urethra is surrounded by a network of blood channels or spaces. When these channels are engorged with blood, the penis becomes erect, that is, an **erection** occurs. Most injuries of the urethra occur at the junction of the membranous and spongy parts. When a metal rod, or **sound,** is passed from the external urethral opening along the urethra, care must be taken to navigate the right-angled turn properly to avoid puncturing the posterior wall of the urethra at this point. In shearing injuries to this area, such as falling astride a fence rail, tearing of the urethra, between the fixed membranous part and the more mobile spony part, may occur, resulting in leakage of urine into the tissues.

Prostate. A mechanism of urinary obstruction produced by prostatic hypertrophy is shown in Fig. 142, which is a sagittal section. The blad-

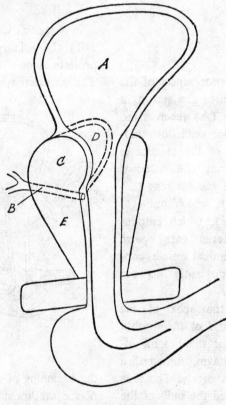

Fig. 142

der cavity (A) is normally constricted just before emptying into the prostatic urethra. The ejaculatory duct (B) is separating the middle lobe (C) of the prostate from the posterior lobe (E). Enlargement of the middle lobe, shown by (D), causes that part of the prostate to project into the normally narrow neck of the bladder, almost obliterating the channel. The enlargement of the prostate may be removed either by entering the bladder through the anterior abdominal wall or by passing a tubelike instrument along the urethra and cutting away small portions. The cutting is done under observation through the same tube.

In Fig. 143 the urinary badder and prostate are seen in a frontal section. The posterior half of both pelvic organs is illustrated. On the pos-

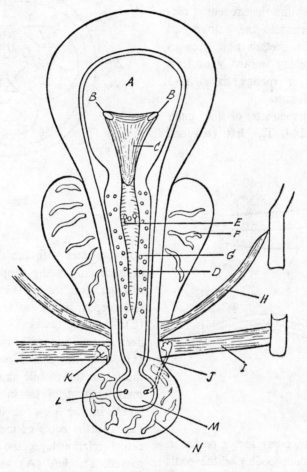

Fig. 143

terior wall of the bladder cavity (A) are two slitlike openings (B), the **ureteral** orifices. The openings are found at the two corners of a triangular area (C) called the **trigone**. It is visible to the naked eye and is formed partly by the extension of the muscle fibers from the ureters into the bladder wall. The apex of the trigone lies at the ureteral orifice. Extending inferiorly along the posterior wall of the prostatic urethra is an elevated ridge (D) called the **urethral crest**. At the highest point of the crest the two orifices of the ejaculatory ducts (E) lie on either side of a nonfunctioning blind pouch called the **utricle**. The prostatic glands (F) open through multiple small ducts (G) into both gutters on either side of the urethral crest. The prostatic glands in addition to the seminal vesicles add their contribution to the seminal fluid.

The prostate rests on the **pelvic diaphragm (H)**, which will be described later. Inferior to the pelvic diaphragm is the **urogenital diaphragm (I)**, through which passes the membranous part of the urethra (J). The paired **bulbo-urethral glands** (K) are located within the urogenital diaphragm; their ducts open into the bulbar part of the urethra (L). The urethra now changes direction (M) as if coming toward you. It is surrounded by erectile or spongy tissue (N) called the **corpus spongiosum.**

Penis. The three components of the penis are illustrated in Fig. 144. The left (A) and

the urethra by traveling in the substance of the corpus spongiosum.

A cross section of the penis is shown in Fig. 145. The left corpus cavernosum (A) partially

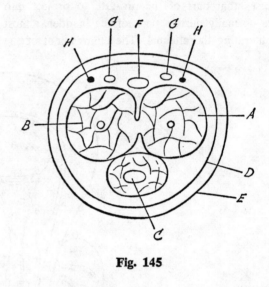

Fig. 145

communicates with the right (B). In the center of each corpus the deep artery of the penis is situated. Inferior to the corpora cavernosum is the corpus spongiosum, containing the urethra (C). The deep fascia (D) of the penis encloses the deep dorsal vein (F), the paired dorsal arteries (G), and the dorsal nerves (H). External to the fascia the mobile skin (E) encloses a minimum amount of fat in the loose subcutaneous layer. In the view of the perineum shown in Fig. 146 the roots of the penis are seen in their actual relationships; the scrotum has been removed. The left (A) and right (B) crura are attached to the conjoined rami, while the bulb of the urethra (C) is attached to the urogenital diaphragm (D). The anus (E) is indicated for purposes of orientation.

Inguinal Region. The anatomy of the inguinal region may be best visualized by following the sequence of events that accompanies the descent of the **testes** from the abdominal cavity to the **scrotum.** In the fetus the testes first appear

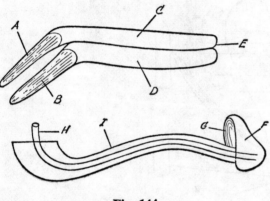

Fig. 144

right (B) **crura** (singular, crus) secure or fasten the corresponding cavernous bodies of the penis (C and D) to the conjoined rami of the hip bone. The bodies, or **corpora cavernosum**, end as a rounded blunt point (E) that fits into the concavity (G) of the glans penis (F). The proper position may be visualized by lowering the bodies in the diagram to sit astride the corpus spongiosum (I) with the bulb of the urethra fitting between the crura. The membranous part of the urethra (H) becomes the penile part of

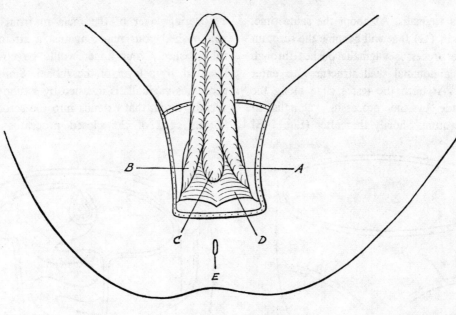

Fig. 146

near the kidney region (Fig. 147). Between the parietal peritoneum (A) and the more external body wall structures (B and F) a testis (C) is shown with the vas deferens (D) extending into the pelvis. The structures represented by B include the three muscle layers as well as the transversalis fascia of the anterior abdominal wall. The skin and superficial fascia are indicated by F. As the testes descend external to the peritoneum, a tubelike extension of the peritoneum (E) extends toward the anterior abdominal wall. This pouch of peritoneum is called

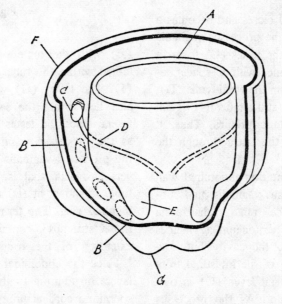

Fig. 147

the **processus vaginalis.** At about the same time, a pouch of skin (G) that will become the scrotum appears. The processus vaginalis pushes through the anterior abdominal wall structures to enter the scrotum. At birth the testes slide along the pathway made by the processus vaginalis to reach the scrotum. Shortly thereafter (Fig. 148)

a covering layer of the facia in front of it. If you pushed your finger against a stretched thin rubber sheet, your finger would be covered, or invested, by a layer of the rubber. Similarly, the processus vaginalis is invested by a tube of transversalis fascia that extends into the scrotum. The fibrous cord of the closed processus vaginalis

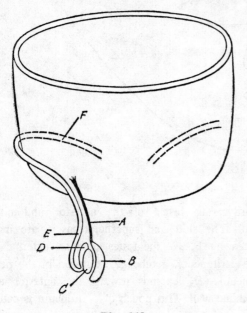

Fig. 148

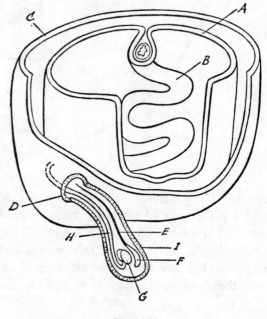

Fig. 149

the processus vaginalis (A) closes and becomes a fibrous cord. That part of the processus vaginalis (B) partially surrounding the testes (C) forms a double-walled coat, or tunic, with a space between the walls. Arising from the **epididymis** (D), the vas deferens (E and F) still retains its original connection with the prostatic urethra. Thus, it marks the route taken by the testes through the abdominal wall.

The first layer of the anterior abdominal wall encountered by the processus vaginalis, and subsequently the testes, is the transversalis fascia (Fig. 149). The parietal peritoneum (A) lines the cavity. Projecting into the cavity but covered by peritoneum is part of the intestinal tract (B). The transversalis fascial layer (C) is outside the parietal peritoneum. As the processus vaginalis meets the transversalis fascia, it pushes

(E) with the tunic of the testes (F) is contained within the tube of the transversalis fascia (I). The testes (G) with its vas deferens (H) also lies within the same tube of transversalis fascia, since the testes follow the same pathway as the processus vaginalis. The point at which the processus vaginalis pushed against the transversalis fascia, that is, the mouth of the fascial tube, is shown at (D) and is called the **internal inguinal ring.** The term **spermatic cord** includes those structures contained within the tubelike extension of the transversalis facia.

Since the abdominal wall consists of muscular layers in addition to the transversalis fascia, the spermatic cord acquires additional coats derived from them (Fig. 150). The layer immediately external to the transversalis fascia, the **transversus abdominis muscle,** does not extend far enough inferiorly to be a barrier to the processus vaginalis. Thus the spermatic cord is not invested by any layer derived from the transversus muscle. The investing layer external to the transversalis fascial coat (A) is derived from the internal oblique muscle and is called the

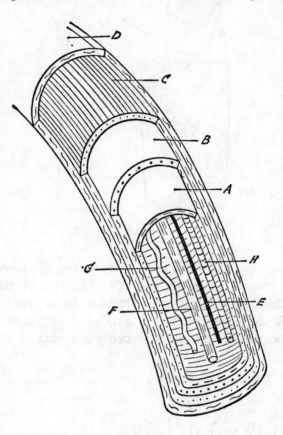

Fig. 150

cremaster muscle layer (B). By pulling on the spermatic cord, it draws the testes closer to the abdominal wall. The fascial layer (C) is derived from the external oblique muscle. The skin and superficial fascial layer are represented by D. The components of the spermatic cord include the vas deferens (F), the fibrous remains of the processus vaginalis (E), and the **spermatic vein** (G), which at this point is really a plexus of veins called the **pampiniform plexus.** When these veins become varicose, the condition is called **varicocele.** The testicular artery (H) supplying the testes is accompanied by sympathetic nerve fibers. Lymph vessels draining the testes also travel in the cord to the posterior abdominal wall.

The pathway through the anterior abdominal wall is called the **inguinal canal** (Fig. 151). The internal inguinal ring is indicated at (A). The spermatic cord (B) traveling in the inguinal canal exits through the external ring (C), an opening in the aponeurosis of the external oblique. The cord (D) continues into the scrotum (E). The inguinal canal and its rings lie above the inguinal ligament, which stretches between the anterior superior iliac spine (G) and the pubis (F).

Inguinal Hernia. If the processus vaginalis remains open after birth, a channel will exist be-

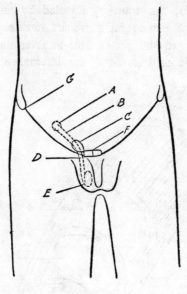

Fig. 151

tween the abdominal cavity and the scrotum. The existence of such a channel is the most frequent cause of inguinal hernia in children and young adults. In Fig. 152 the open processus vaginalis (A) is filled by the intestine (B), which extends from the abdominal cavity to the scrotum (C). This type of hernia is called an indirect inguinal hernia and is corrected basically by pushing the intestine back into the abdominal cavity and tying off the processus vaginalis.

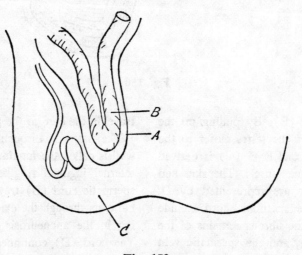

Fig. 152

THE UTERUS

The uterus and its major components are shown in Fig. 153. The **fundus** (A) of the uterus is that part above the level of the **uterine tubes** (E). The neck of the uterus is the **cervix** (C), which penetrates the wall of the **vagina** (D). The largest segment of the uterus is the body

(B). The length of the uterus in an adult female is about three inches; the fundus and body are two inches in length and the cervix accounts for the lower inch. The **uterine,** or **Fallopian, tubes** end in fingerlike processes called **fimbria**

(F), which help to attach the tube to the **ovary** (G). The **ovarian ligament** (H) secures the ovary to the uterus, while the round ligament (I) of the uterus travels to the anterior abdominal wall through the inguinal canal and ends

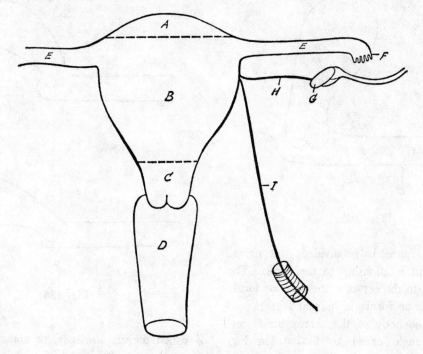

Fig. 153

in the **labia majora.** These are the larger lips of the **vulva,** the external genital organ of the female. Although the round ligament is not the female counterpart of the vas deferens, it follows the same path through part of its course.

The uterus is bisected in Fig. 154 showing its cavity (A), which is continuous with the hollow uterine tubes (B). The **cervical canal** (C) terminates at the external **os,** or orifice (D), to open into the vagina. Thus, there is a direct communication between the external environment and the peritoneal cavity in the female. No similar communication exists in the male.

The normal relationship of the uterus (D) with the vagina (A) is shown in Fig. 155. The vagina is approximately 3½ inches long but is capable of much greater distension. The long

axis of the vagina makes an angle of approximately forty-five degrees with the floor in the standing female. The anterior wall of the vagina (B) is shorter than its posterior wall (C). The cervix of the uterus projects through the anterior

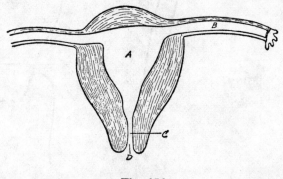

Fig. 154

wall of the vagina at right angles to its long axis. The space between the wall of the vagina and the intruding cervix is called the **fornix.** The fornix thus encircles the cervix. For practical

Fig. 156B, where the cervix lies more in line with the long axis of the vagina. The uterus is called **retroverted** and is more likely to slide down the vagina. If this happens, the condition

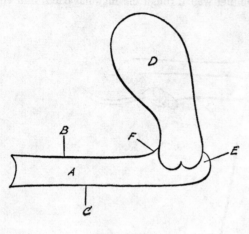

Fig. 155

purposes the space is subdivided and named according to its relationship to the cervix. The space anterior to the cervix is the **anterior fornix** (F) the **posterior fornix** is indicated at E.

Normally, the body of the uterus bends anteriorly, the angle called **anteflexion.** In Fig. 156A the uterus is retroflexed, or bent backward. A more abnormal position is shown in

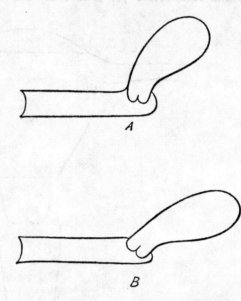

Fig. 156

is called uterine prolapse, or sometimes "fallen womb."

The broad ligament of the uterus (Fig. 157)

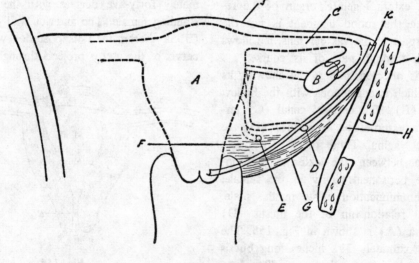

Fig. 157

is formed by a sheet of peritoneum (J) draped over the structures attached to the side walls of the uterus (A). The structures contained within the broad ligament are the uterine tube (C), the ovarian ligament, and the uterine artery (D), which supplies blood to the uterus.

Ovary. The ovary (B) is not completely enclosed in the peritoneum of the broad ligament. Thus, when ova are extruded from its surface, they may fall into the peritoneal cavity. The attachment of the fimbria to the ovary usually insures that the ovum is directed into the uterine tube and thence to the uterine cavity. The broad ligament extends from the side wall of the uterus to the side wall of the pelvis, which is formed by the **pelvic diaphragm** (G), the **obturator internus muscle** (H), and the hip bone (I). K represents a cut edge of the broad ligament. The broad ligament contributes little to the support of the uterus. A strong support of the uterus is the **cardinal ligament** (F) extending from the cervix and vagina to the side wall of the pelvis. The relationship of the ureter (E) to the cervix, uterine artery, and the cardinal ligaments necessitates caution during surgical removal of the uterus.

A sagittal section through the broad ligament (Fig. 158) shows the anterior wall (A) of the peritoneal fold as well as the posterior wall (B). Between the two peritoneal folds are the uterine tube (C), the ovarian ligament (D), and the round ligament (F). Near the base of the folds is the uterine artery (G) on its way to the uterus. Projecting into the base of the broad ligament is the cardinal ligament (H), which rests on the curved floor and side wall formed by the muscular pelvic diaphragm (I). The attachment of the ovary to the posterior wall of the broad ligament leaves a surface free of peritoneum (E), from which the ova are extruded. We would not ordinarily expect to find both an ovary and an ovarian ligament in the same section.

Supports of Uterus. By removing the major part of the pelvic organs (Fig. 159) and looking

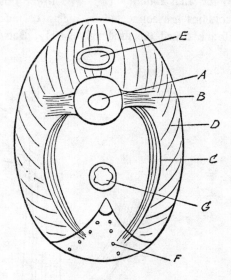

Fig. 159

down into the basinlike floor, the chief supports of the uterus are seen. The cardinal ligament (B), extending from the cervix (A), was described previously. Curving postero-laterally from the cervix, the utero-sacral ligaments (C) extend to the anterior surface of the **sacrum** (F), encircling the rectum (G). Anterior to the cervix is the bladder (E). The muscular floor of the pelvis (D) is called the pelvic diaphragm. By inserting into the walls of the urethra, vagina,

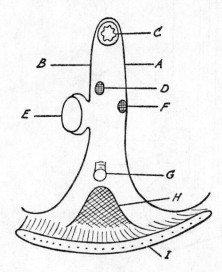

Fig. 158

and rectum, this muscular diaphragm provides strong support to the pelvic organs.

RECTUM AND ANAL CANAL

The lower end of the intestinal tract is shown in Fig. 160. The tube is lined with mucous membrane (A) surrounded by an inner circular layer of muscle (B), which is in turn covered by a longitudinal muscle layer (C). Inserting into the wall of the tube and giving strong support is the muscular pelvic diaphragm (D). The lower end of the tube is surrounded by bands of circular muscle called the **external anal sphincter** (E). Protruding into the lumen of this canal are eight to ten small columns of mucous membrane called the **anal columns** (F). The lower tips of the columns are connected by a flap of mucous membrane called the **anal valves** (H). Between

adjacent columns there is a valley, or gutter (G). Where the inferior end of the gutter meets the anal valve a small pouch called the **anal sinus**, or **crypt**, is formed. This area is often the site of damage resulting from the passage of hard feces. The valve may be torn, producing a break or fissure in the mucous membrane. Infection in the anal sinus may spread through the mucous membrane as well as the muscular walls of the canal to form an abscess in the fat surrounding the external surface of this canal. The channel connecting the abscess in the external fat with the anal sinus is called a **fistula**.

The serrated or scalloped margin formed by the lower ends of the anal columns is called the **pectinate line.** Pectinate means comblike, that is, resembling the teeth of a comb. Surgeons usually use this line to mark the junction between the rectum above and the anal canal below. The area J is covered by a skinlike lining, while the

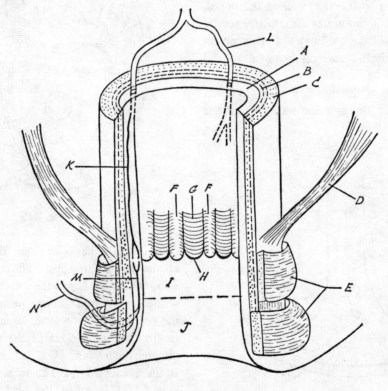

Fig. 160

area I is lined with a type of cell midway between the flat cells of the skin and the tall, columnar cells of the rectal mucosa. This area is called the **pecten.** It is a useful landmark since it indicates a type of "watershed" for the nerves, blood vessels, and lymphatics of the canal. Above the pecten the mucous membrane is supplied by nerves of the autonomic system; thus, the membrane can be cut or burned without producing pain. Below the pecten the lining is highly sensitive. Early cancer of the rectum is therefore painless, while cancer of the anal canal is painful. Lymph vessels above the pecten drain into lymph nodes in the pelvis, and those from the anal canal drain into the superficial nodes in the groin, the inguinal nodes. Within the loose mucous membrane the veins form a network, or plexus.

Those above the pecten (K) drain into the **superior rectal veins** (L), which are part of the portal system. The plexus of veins inferior to the pecten (M) drain into the **inferior rectal vein** (N), which drains into the caval system. They eventually reach the inferior vena cava. In the region of the pecten, the two venous systems communicate. Increased venous pressure in the portal venous system may produce varicosities or hemorrhoids in this area, although this is not their only cause. Varicosity of veins above the pecten is called internal hemorrhoids, while the external hemorrhoids are caused by varicosities of those veins below the pecten.

The arterial blood supply to the pelvic organs is shown in Fig. 161. The common iliac artery (A) arising from the bifurcation of the aorta

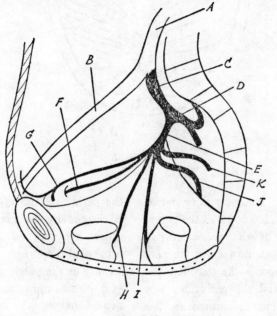

Fig. 161

again divides into the external iliac artery (B), which goes to the lower limb, and the internal iliac (C), which supplies pelvic and perineal structures. The internal iliac gives off a posterior trunk (D), which supplies the posterior wall of the pelvis as well as the **gluteal** region. The anterior trunk (E) gives rise to the **umbilical** artery (G), which supplies the bladder and then continues on the anterior abdominal wall as a solid cord. This artery in fetal life carried blood to the placenta, so after birth the lumen of the artery becomes obliterated. Other branches of the anterior trunk are the obturator (F), the inferior vesicle (H) to the bladder, the middle rectal (I), the **internal pudendal** (J) to the perineum, and the **inferior gluteal** (K) to the region of the same name.

Pelvic Diaphragm. In order to appreciate the basic plan of the pelvis and perineum, first review the bony walls of the pelvis. In Fig. 162 you are looking from above into the pelvic cavity. The anterior part is indicated by the pubic symphysis (A); posteriorly is the sacrum (B). The floor of this cavity is concave and is formed by large flat muscles stretching from the side walls of the pelvis (C) to meet and join each

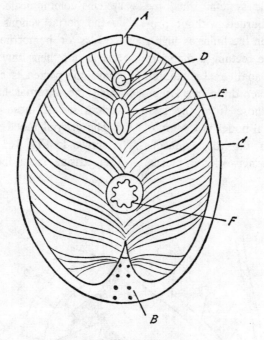

Fig. 162

other in the midline. This muscular floor, while formed by several muscles, is known as the **pelvic diaphragm.** The organs of the pelvic cavity, which pierce this diaphragm, now enter the perineum; the pelvic diaphragm is the partition between the pelvic cavity and the perineum. The organs that pierce the pelvic diaphragm are the urethra (D), the vagina (E), and the rectum (F). The muscle fibers of the pelvic diaphragm insert into the walls of these tubes, and the muscle tone restrains their inferior movement.

In a sagittal section through the pelvis and perineum (Fig. 163), the concave surface of the pelvic diaphragm (A) forms the floor of the pelvic cavity. Inferior to the pelvic diaphragm is a smaller reinforcing muscular diaphragm in the perineum, the urogenital diaphragm (B). It is pierced by the urethra, and from its inferior aspect the root of the penis is attached. The urogenital diaphragm does not extend so far posteriorly as the pelvic diaphragm; thus, the rectum (C) pierces only the pelvic diaphragm. The **pudendal nerve** (G) and the **internal pudendal artery** (I) arise within the pelvic cavity. They skirt the pelvic diaphragm by going out through the **greater sciatic foramen,** turning around the **ischial spine,** and entering the perineum through the **lesser sciatic foramen.** The pudendal nerve (H) branches to supply the anal canal, the skin, muscles of the urogenital diaphragm, and the penis. Arterial branches essentially follow the nerves. The deep dorsal vein of the penis (J) enters the pelvic cavity through a gap between the pubic symphysis and the anterior edge of the urogenital diaphragm. A membranous layer of superficial fascia (D) is attached to the posterior border of the urogenital diaphragm and runs anteriorly to surround the scrotum (E), where it contains muscle fibers. These muscle fibers constitute the **dartos muscle.** Contraction of this muscle causes wrinkling of the scrotum when the scrotum is exposed to cold. This membranous layer forms a tube around the penis and extends into the anterior surface of the abdominal wall (F). If the urethra is torn as it emerges from the urogenital diaphragm, the escaping urine can spread from the perineum into the scrotum, along the penis, and up the

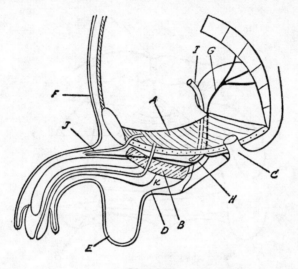

Fig. 163

anterior abdominal wall, that is, within the space (K) lined by this membranous layer of superficial fascia.

A frontal section through the pelvis and perineum (Fig. 164) would show the same structures, but in a plane at right angles to Fig. 163. Lining the bony wall of the pelvis (A) is the obturator internus muscle (B). Arising from the side wall is the muscular pelvic diaphragm (C), which supports the bladder and prostate. Inferior to the pelvic diaphragm is the perineal structure, the muscular urogenital diaphragm (D), traversed by the membranous portion of the urethra. The membranous layer of superficial fascia (E) is attached laterally to the conjoined rami. The space between this layer (E) and the urogenital diaphragm is called the superficial perineal pouch (F). This is the same space where urine from a ruptured urethra collects. It will be observed that the crura of the corpora cav-

ernosum (H), as well as the bulb of the urethra (G), are attached to the inferior surface of the urogenital diaphragm.

The inferior surface of the urogenital dia-

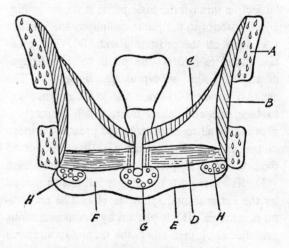

Fig. 164

phragm (C, Fig. 165) is triangular in outline. The muscles are attached laterally to the conjoined rami. It has a free, or unattached, posterior margin. The pubic symphysis (A) and the ischial tuberosities (B) are shown for orientation. The muscle of the urogenital diaphragm where it surrounds the urethra (D) is arranged in a circular manner to form the external sphincter of the urethra. The gap, or space, between the pubic symphysis and the anterior margin of the diaphragm (E) is occupied by the deep dorsal vein draining the penis (or **clitoris** in the female). The vein enters the pelvic cavity through this opening.

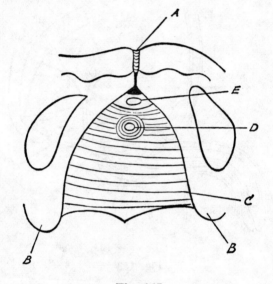

Fig. 165

Male Pelvis. In order to review the relationships of the pelvic and perineal structures a sagittal section through the male pelvis is shown in Fig. 166. Posterior to the pubic symphysis the bladder (A) rests on the prostate gland (B). The prostate gland is held to the pubis by paired ligaments (J), the **pubo-prostatic ligaments.** The seminal vesicle (C) joins the vas deferens (D) to form the ejaculatory duct, which pierces the prostate gland to empty into the prostatic portion of the urethra. The rectum (E) lies anterior to the sacrum and pierces the pelvic diaphragm (G). Its lower part, the anal canal, is surrounded by the external anal sphincter (F). The urogenital diaphragm (H) is pierced by the membranous urethra, which then enters the corpus spongiosum of the penis (I). The parietal peritoneum lining the anterior abdominal wall is reflected onto the superior surface of the bladder (K) and dips down between the bladder and the rectum but does not reach the pelvic floor. The lower part of the rectum is thus not covered by peritoneum. Between the peritoneum and the pelvic floor a tough fibrous partition (L) separates the prostate from the rectum. It is believed that this fibrous wall prevents for a time a cancer of the prostate from spreading directly to the wall of the rectum.

Female Pelvis. In a diagram (Fig. 167) of a similar section through the female pelvis, the relationships should be compared with those found in the male pelvis. The urinary bladder (A) is in close relationship to the uterus (B). Enlargement of the pregnant uterus may press

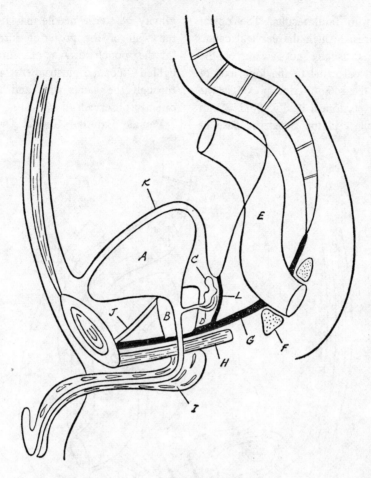

Fig. 166

on the bladder, decreasing its capacity and causing frequency of urination. The female urethra is in close contact with the anterior wall of the vagina (C). The passage of the baby's head along the vagina during labor stretches not only the vagina but also the urethra. Pressure by the fetal head against the lower margin of the symphysis compresses and injures the urethra. For these reasons some women are unable to urinate voluntarily for a short time after delivery. Both the vagina and the urethra pierce the pelvic (F) and urogenital (G) diaphragms. The rectum (D) and external anal sphincter (E) are similar to those of the male.

The reflection of the peritoneum from the anterior abdominal wall onto the upper surface of the bladder dips for a short distance between the bladder and uterus. This peritoneal layer continues over the uterus (H), completely covering its posterior surface and even covering the posterior fornix of the vagina (I). It dips between the posterior fornix and the rectum to form the **recto-uterine pouch** (J) before continuing on to the rectum.

The intimate relationship of the peritoneum to the posterior fornix has been the cause of tragedy as well as the source of life-saving information. The introduction of a pointed (or even a blunt) instrument, such as a knitting needle, along the vagina for the purpose of producing an abortion has resulted in a rupture of the posterior fornix, carrying infection into the peritoneal cavity with fatal results. The object is to enter the uterus through the cervical canal, but the amateur is usually not aware of the relation of the cervical canal to the long axis of the vagina. Since the recto-uterine pouch is the lowest part of the peritoneal cavity, fluid within that cavity will collect in the pouch because of gravity. A sterile needle may be inserted through the wall of the posterior fornix and the fluid in the pouch drawn off for examination. A lighted tubelike instrument may be inserted through the same wall and the pelvic organs directly observed.

Female External Genital Organs. The exter-

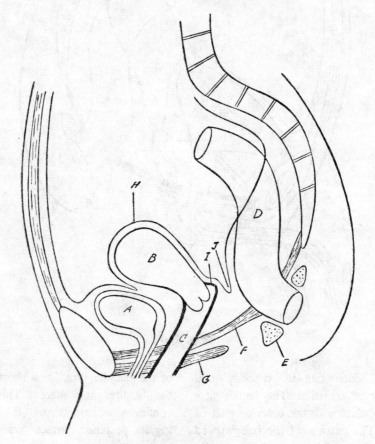

Fig. 167

nal genital organs of the female are shown in Fig. 168. The labia majora (A) are pulled apart to expose the **labia minora,** the lesser lips (B). Each lip splits to enclose the clitoris (C), forming its foreskin. The area between the labia minora (D) is called the **vestibule,** for it is the anteroom from which two passageways are entered. The more anterior opening in the vestibule is the urethral opening (E), while the larger, posterior one is the vaginal opening (F). Paired glands called the **greater vestibular glands** each open into the vestibule via a duct (G). These glands, while providing lubrication, are frequently infected by venereal disease. The anal opening is shown at (H).

On a slightly deeper level after the skin has been removed (Fig. 169), the bulbs of the vestibule (A) lie deep to the labia majora against

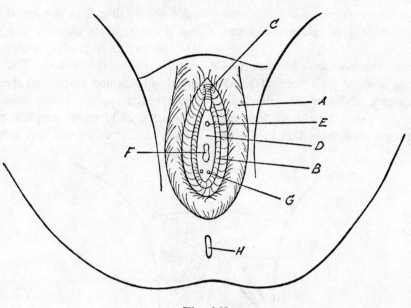

Fig. 168

the lateral walls of the vestibule. They contain a varying amount of erectile tissue. At their posterior margin are found the greater vestibular glands (B), which open into the vestibule. Arising from the conjoined rami, as well as from the urogenital diaphragm (G), are the two crura (C) of the corpora cavernosum of the clitoris, which later merge to form the body of the clitoris. The labia minora (D) enclose the vestibule into which the vagina (E) and the urethra (F) open. The pelvic diaphragm is shown at (H). The anal canal is surrounded by the external anal

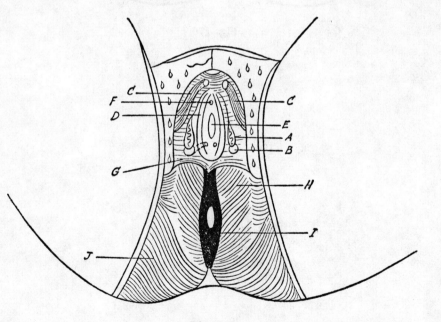

Fig. 169

sphincter (I). More posteriorly, the major muscle of the gluteal region, the **gluteus maximus,** is indicated by (J).

In Fig. 170 the two corpora cavernosa (A) unite to form the body of the clitoris (B). Although a few strands of tissue from the bulbs of the vestibule (C) reach the clitoris, they are of minor importance. Remember that in the fe-male the urethra does not travel in the clitoris but has a separate opening. The vaginal orifice (D) in virgins is partially closed by a fold of mucosa called the **hymen.** The hymen varies greatly in size and shape, and the most frequent cause of its rupture is intercourse. During the birth of a child more complete rupture occurs, so that finally only small tags remain.

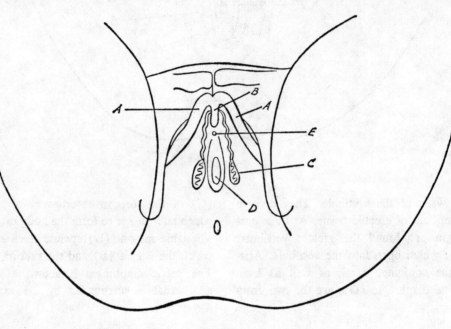

Fig. 170

CHAPTER 8

THE UPPER EXTREMITY

Certain movements of the upper extremity must first be defined if the function of the various muscles is to be clearly understood. The upper extremity, or limb (Fig. 171), consists of the

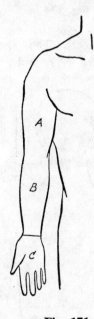

Fig. 171

arm (A), which extends from the shoulder to the elbow, and the **forearm** (B), which extends from the elbow to the wrist. Distal to the wrist is the **hand** (C), the remaining part of the limb.

MOVEMENTS OF THE ARM, SHOULDER, FOREARM, HAND

Arm Movements. One movement of the arm is shown in Fig. 172. From the starting position (A), the arm is brought forward (B) and upward (C) toward the side of the head. This is

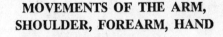

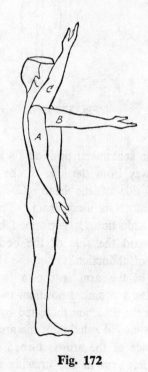

Fig. 172

flexion of the arm. Movement is taking place between the **humerus,** or arm bone, and the shoulder joint. The opposite movement, **extension** of the arm, starts at the end of flexion (C), with the arm brought back to the starting position (A). Abduction of the arm (Fig. 173)

rotated; when it is turned in the direction of the arrow (B), **lateral rotation** of the arm will be performed.

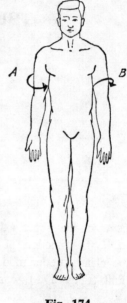

Fig. 174

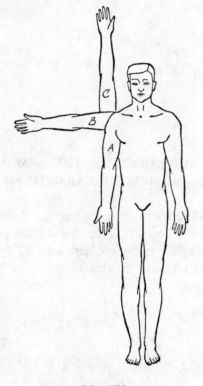

Fig. 173

begins at the anatomical position (A). The arm is moved away from the side of the body (B) and to the side of the head (C). The word *abduct* means "lead away from." The opposite movement, adduction, is performed by bringing the arm toward the side of the body; that is, the reverse of abduction.

Rotation of the arm is shown in Fig. 174. Starting in the anatomical position, the right arm is rotated in the direction indicated by the arrow (A). This is **medial rotation** of the arm, in which the lateral side of the arm is turned toward the body. The left arm in the drawing is medially

Shoulder Movements. The shoulder (Fig. 175) can be **elevated** (A), as in shrugging, or

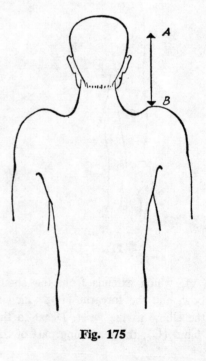

Fig. 175

moved in the opposite direction (B), called **depression.** The shoulder can also (Fig. 176) be

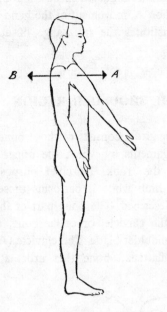

Fig. 176

moved forward (A), called **protrusion,** as well as moved backward (B), called **retraction.**

Movements of the Forearm. These are more

Fig. 177

restricted. Starting in the anatomical position (A in Fig. 177), the forearm is flexed at (B) and more acutely flexed at (C). To extend the forearm, begin at position (C) and bring the forearm to the original position (A). The elbow joint will not permit abduction or adduction of the forearm. Flex the forearm with the elbow held tightly against the side of the body and the palm facing upward (Fig. 178A). The forearm is now

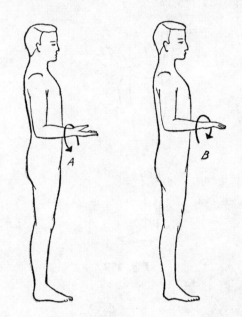

Fig. 178

supinated. **Pronation** of the forearm is performed by turning the hand so that the palm faces the floor (Fig. 178B). It may be helpful to remember that when a person falls in a prone position his forearm will be pronated.

Observe the hand during pronation and supination and note that it turns through 180 degrees (two right angles), the normal range of movement. Release the elbow from the side of the body and then observe the range of supination and pronation: the hand now turns through slightly more than three right angles. Why the greater range of movement when the elbow is not held against the body? Rotation of the arm adds 90 degrees of movement to the normal

180 degrees of supination or pronation of the forearm. When testing for supination or pronation, one must hold the elbow to the side of the body to eliminate rotation of the arm.

Hand Movements. Flexion and extension of the hand are shown in Fig. 179. Starting with

Fig. 179

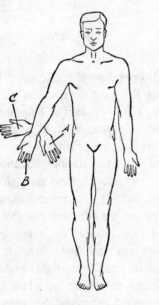

Fig. 180

position B, full flexion is position A; extension is position C. In Fig. 180 the hand is adducted in position A and abducted in position C. Beginning at position A, movement of the hand through B to C constitutes the full range of abduction of the hand.

THE SHOULDER REGION

The **shoulder girdle** consists of those bones, muscles, and ligaments by which the upper limb is attached to the trunk. Its chief purpose is to support the limb while it performs those movements just described. The bony part of the girdle consists of the **clavicle,** or collar bone, and the **scapula,** or shoulder blade. The clavicle (A in Fig. 181) is a flattened bone that articulates with

Fig. 181

the **sternum,** or breast bone (E), at its medial end (B) and with a cromial process (D) of the scapula (F) at its lateral end (C). The medial half of the clavicle curves forward; this curvature can readily be felt in a living person. G represents the arm bone, the humerus. The remaining part of the bone girdle is the scapula (Fig. 182), shown in three different views. The prom-

as muscles are attached to this process. Inferior to the acromial process (E) is a shallow cuplike depression (G) called the **glenoid fossa.** Into this fossa fits the head of the humerus. Two borders or edges of the scapula should be observed: the **vertebral border** (I), which lies near the vertebral column, and the **axillary border** (H), which lies near the posterior fold of the **axilla,** or armpit. The superior angle (K) and the inferior angle (J) are useful landmarks because they can readily be felt.

In the bony girdle, seen from above and behind (Fig. 183), the medial end of the clavicle

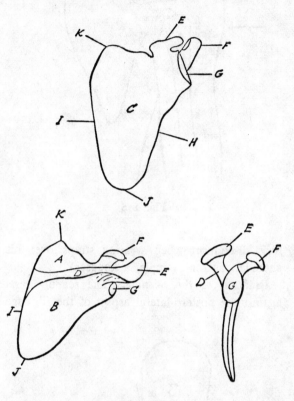

Fig. 182

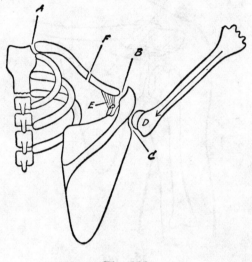

Fig. 183

inent feature on the posterior surface is the **scapular spine** (D), which separates two shallow depressions, or fossae. The **supraspinatus fossa** (A) and the **infraspinatus fossa** (B) are areas from which arise muscles bearing the same name. Remember that *supra* means "above" and *infra,* "below." The spine ends laterally as a free blunt end (E), called the **acromial process.** The area C, seen from the anterior aspect, is called the **subscapularis fossa;** from this area the **subscapularis muscle** arises. From the superior edge of the scapula a fingerlike process (F) points in an anterior as well as a lateral direction and is called the **coracoid process.** Ligaments as well

articulates with the sternum, forming the **sternoclavicular joint** (A). The lateral end of the clavicle articulates with the acromial process of the scapula, forming the **acromio-clavicular joint** (B). (Most joints are named by compounding the two bones that form the joint.) The head of the humerus (D) articulates with the glenoid fossa to form the **gleno-humeral joint** (C). An important ligament (E) that ties the coracoid process to the clavicle is the **coraco-clavicular ligament.** As one falls on the outstretched hand, the resulting force travels up the humerus in the direction of the arrow to the glenoid fossa. Some of this force is then transmitted through the coraco-clavicular ligament to the clavicle and thence to the sternum, which is part of the stable framework of the trunk. Sometimes the force is excessive and a fracture of the clavicle occurs (F). Since the clavicle is like a strut pushing the scapula laterally, a fracture of the clavicle permits the shoulder to fall downward, forward, and in a medial direction.

The acromio-clavicular joint (Fig. 184) is fre-

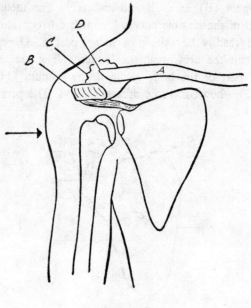

Fig. 185

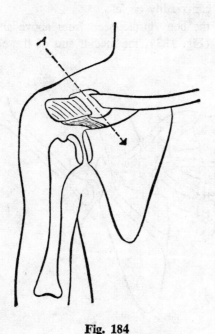

Fig. 184

quently the site of injury, especially in males participating in a contact sport. The type of injury is apparently related to the direction of the joint line (A) that passes through the middle of the joint. Observe that the joint line points in a medial as well as downward direction. Forceful contact of the shoulder as indicated by the arrow (Fig. 185) may cause the clavicle (A) to override the acromial process (B), resulting in partial tearing of the **joint capsule** (C). The now free end of the clavicle (D) may readily be seen as well as felt. When pressure is applied to the clavicle, it moves in an inferior direction. Although the victim is not incapacitated, he often

feels discomfort when wearing straps over his shoulder, as for example, a knapsack.

Rotation of the Scapula. The scapula rests against the postero-lateral aspect of the rib cage

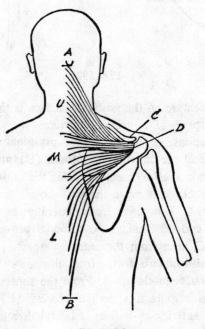

Fig. 186

(thorax), and muscles are solely responsible for maintaining its position. The **trapezius muscle** (Fig. 186) is one important in supplying support. Its origin extends from the **external occipital protuberance** (A), which is the most prominent knob at the back of the skull, to the bony prominence, or spine, of the twelfth thoracic vertebra (B). It is inserted into the clavicle (C), the acromion (D), and the spine of the scapula. The trapezius muscle is so massive that only a part of it may be used on occasion. Contraction of the upper part (U) will elevate or shrug the shoulder; the middle third (M) will retract the shoulder; and the lower part (L) will depress the shoulder. All parts of the muscle acting together will rotate the scapula (Fig. 187).

powerful muscle that assists the trapezius in rotating the scapula. In this diagram the scapula is pulled away from the rib cage so that more of

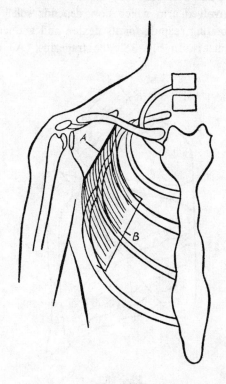

Fig. 188

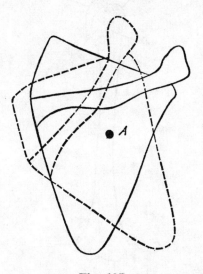

Fig. 187

Rotation of the scapula may be considered as movement around an imaginary horizontal axis through the point indicated at A. The rotated position of the scapula is shown by the interrupted lines and will be called **upward rotation.** Place your hand over a person's scapula and ask him to abduct his arm. You should be able to feel the scapula rotate upward. When he adducts the arm, **downward rotation** of the scapula occurs.

The **serratus anterior muscle** (Fig. 188) is a

the muscle can be seen. The serratus anterior arises from the upper two-thirds of the rib cage (B) and is inserted into the vertebral border of the scapula (A). The lower part of the muscle is much thicker and therefore more powerful than the upper half. The muscle will pull the scapula forward in protrusion of the shoulder, and its lower fibers will rotate the scapula upward during abduction as well as flexion of the arm. During abduction or flexion the arm has a normal range of movement of approximately 180 degrees. One-third of the total movement is caused by rotation of the scapula and two-thirds by the gleno-humeral joint. Paralysis of the serratus anterior or trapezius muscles would, therefore, seriously limit flexion and abduction of the arm. If the gleno-humeral joint is diseased and painful to move, the humerus and the glenoid fossa are sometimes surgically joined so that no movement, and hence no pain, occurs. The patient nevertheless can perform many ordinary activities with the involved arm which now depends solely on the rotating scapula for its flexion and abduction movements. In Fig. 189 the trapezius (A) and the serratus anterior (B) are relaxed when the arm is at the side of the body. During abduction or flexion of the arm (Fig. 190) the two muscles contract and pull in the directions indicated by the arrows to rotate the scapula upward.

THE GLENO-HUMERAL JOINT

The gleno-humeral joint is a synovial joint having those general features of joints already described. Certain features of the humerus important to an understanding of the joint are shown in Fig. 191. The articulating surface of the head

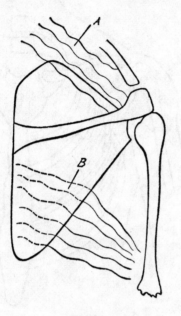

Fig. 189

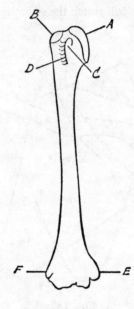

Fig. 191

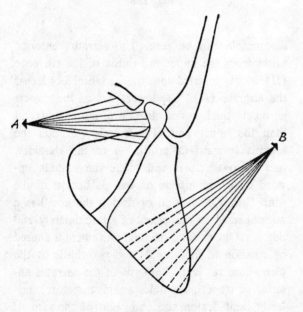

Fig. 190

(A) is covered with articular cartilage. Distal to the head are two bony prominences called tubercles, one larger than the other. The **greater tubercle** (B) faces laterally and the **lesser tubercle** (C) is on the anterior surface of the humerus. Between the two tubercles a shallow groove (D) forms a bed for a muscle tendon. At the distal end of the humerus a bony prominence (E) called the **medial epicondyle** can be felt on the medial side of the elbow. The medial epicondyle always points in the same direction as the head of the humerus, which cannot be felt. The medial epicondyle is therefore a reliable guide to the position of the head of the humerus. The less prominent **lateral epicondyle** (F) may be felt on the lateral side of the elbow. The other bone that helps to form the joint is shown from a lateral view (Fig. 192). The round head of the

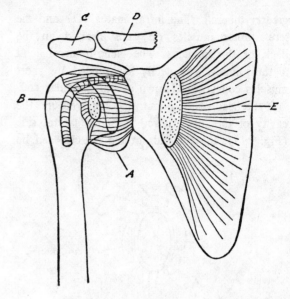

Fig. 193

Fig. 192

humerus fits against the shallow cup of the glenoid fossa (A). The enlarged end of the spine of the scapula, the acromial process (B), forms a partial bony roof over the joint. The coracoid process (C) projects anterior to the joint.

Muscles. In an anterior view of the glenohumeral joint (Fig. 193), a sleeve of fibrous connective tissue (A) forms the capsule of the joint. From the upper end of the glenoid fossa a tendon arises within the fibrous capsule and emerges from the capsule at B to lie in the groove between the tubercles. This tendon is the long head of the **biceps muscle** of the arm. The acromial process (C) is superior to the joint and articulates with the clavicle (D). The subscapularis muscle (E), cut in this diagram to show the capsule, covers the anterior aspect of the joint and inserts onto the lesser tubercle. On the

posterior aspect of the joint (Fig. 194) the **fibrous capsule** (A) is covered by three muscles. A portion of each is removed here to show the

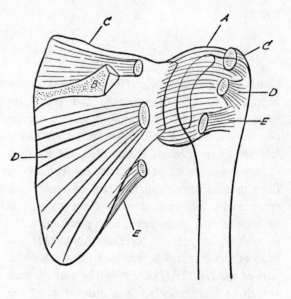

Fig. 194

capsule. The **supraspinatus** (C), above the partially removed scapular spine (B), runs over the superior aspect of the joint to be inserted on the greater tubercle. The infraspinatus (D) and the **teres minor** muscles (E) are inserted on the back of the humerus. The gleno-humeral joint is thus almost completely surrounded by short muscles collectively known as the **muscular cuff.**

If the head of the humerus is removed and the cavity of the joint viewed from the lateral side (Fig. 195), the cut ends of the muscles can be

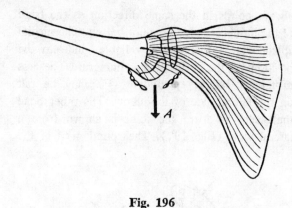

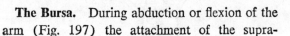

Fig. 196

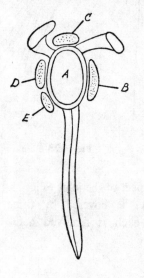

Fig. 195

The Bursa. During abduction or flexion of the arm (Fig. 197) the attachment of the supra-

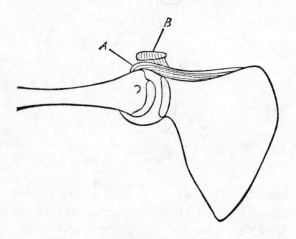

Fig. 197

seen. The subscapularis (B) is anterior; the supraspinatus (C) is superior; and the infraspinatus (D) and teres minor (E) are posterior and surround the glenoid fossa (A). Only in the inferior space between E and B is there a gap of any consequence in the muscular cuff. Since these muscles strengthen the joint by reinforcing the capsule, the weakest part of the capsule is its inferior wall. Sometimes, as a result of a fall in which the arm is forcefully abducted, the head of the humerus is pushed against this weak part of the capsule (Fig. 196). The capsule may stretch or tear, resulting in a **dislocation** of the head of the humerus. A dislocation is the displacement of a bone from its normal position. Although dislocation does not imply a break **(fracture)** of the bone, dislocation and fracture of the same bone may occur.

spinatus muscle (A) comes into close contact with the overhanging acromial process (B). Friction between these two adjacent structures is minimized by a **bursa** (Fig. 198). A bursa is a

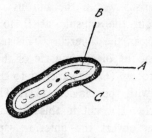

Fig. 198

fibrous bag (A) that is lined with a smooth, glistening membrane (B) and contains a small amount of viscous fluid (C). The position of the bursa is shown in Fig. 199. The bursa (A) lies

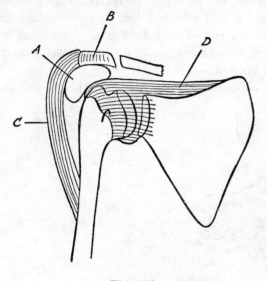

Fig. 199

inferior to the acromial process (B) and extends laterally to separate the deltoid muscle (C) from the humerus and supraspinatus muscle (D). This bursa is named from its position; because it is under the acromion as well as the deltoid muscle it is sometimes called the **subacromial bursa** and at other times the **subdeltoid bursa**. A normal, healthy bursa is not ballooned out as shown here, but is collapsed so that the surfaces of the lining touch. One outside surface of the fibrous bag is firmly attached to the inferior surface of the acromion and the deep surface of the deltoid; the opposite outside surface adheres to the muscular cuff and humerus. When the arm moves in abduction and flexion, only the opposing surfaces of the smooth lining of the bursa rub against each other.

Sometimes, as a result of the accumulated effects of everyday "wear and tear," and at other times for reasons unknown, the bursa becomes swollen and painful. Deposits of calcium may be found in the bursa wall. The resulting pain may render movement at the joint impossi-

ble. This condition is called **bursitis**, meaning "inflammation of the bursa." The supraspinatus tendon just inferior to the bursa may also be involved in the process. The fibers of the tendon become frayed and weakened so that, after a sudden forceful abduction movement of the arm, the tendon may tear across and become separated from the fleshy part of the muscle.

MUSCLES THAT MOVE THE ARM

Deltoid Muscle. The powerful **deltoid** muscle that is draped like a curtain over the gleno-humeral joint is shown in three different views in Fig. 200. The anterior part of the muscle (A) arises

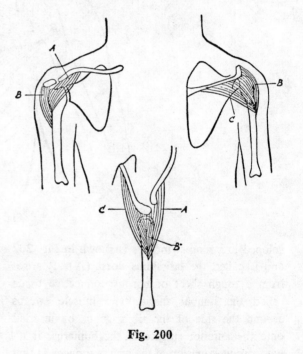

Fig. 200

from the clavicle and can flex the arm; the lateral part of the muscle (B) arises from the acromion and abducts the arm; and the posterior part (C), arising from the scapular spine, can extend and adduct the arm. The deltoid muscle gives the rounded contour to the shoulder; therefore, when it is paralyzed, the shoulder is more square in outline.

Pectoralis Major. From the anterior chest wall there arises a large fan-shaped muscle (Fig. 201) called the **pectoralis major,** which inserts onto the anterior surface of the humerus (C). One part of the muscle (A) arises from the medial half of the clavicle and is known as the **clavicular head.** The larger head arises from the sternum, and its lower edge (D) can be grasped between the fingers in the anterior fold of the axilla. The clavicular head (A) contracts to flex the arm, and the sternal head (B) extends the arm. Both parts will depress the shoulder and adduct the arm. This muscle is frequently removed with a cancerous breast because the breast rests on the muscle, but even with its removal flexion of the arm is still possible. You will recall that the anterior part of the deltoid muscle also flexes the arm; hence movement of the arm, though weaker, is not lost.

Latissimus Dorsi. A muscle that is well de-

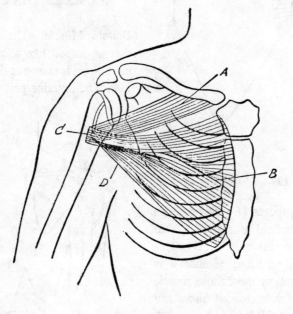

Fig. 201

veloped in strong swimmers is shown in Fig. 202 and is called the **latissimus dorsi** (A). It arises from a tough sheet of fibrous connective tissue called the lumbar fascia. The muscle sweeps around the side of the rib cage to be inserted onto the anterior surface of the humerus. It allows strong extension of the arm, a movement that propels the body through the water. The latissimus dorsi also adducts the arm. It can be felt between the fingers when the posterior axillary skin fold is grasped. Three muscles that hold the scapula in place against the rib cage are the **levator scapula** (B), the **rhomboideus minor** (C), and the **rhomboideus major** (D). The latter two muscles arise from the spinous processes (E) of the vertebrae. What muscle was removed in order to see these three muscles? The trapezius.

Biceps. The most prominent muscle on the anterior aspect of the arm is the biceps, or two-headed, muscle of the arm, shown in Fig. 203. One head, called the **long head** (A), was previously described as arising within the fibrous capsule of the gleno-humeral joint. The **short head** (B), arising from the coracoid process, unites with the long head to form the biceps muscle. The tendon of this muscle is inserted onto the bony prominence of the radius (C), known as the **radial tuberosity.** The biceps muscle both flexes and supinates the forearm.

Triceps. On the dorsal surface of the arm a

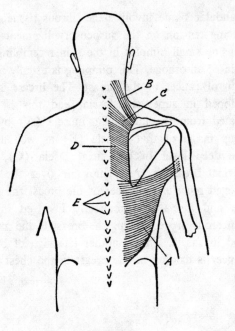

Fig. 202

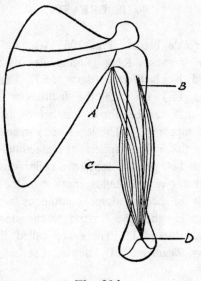

Fig. 204

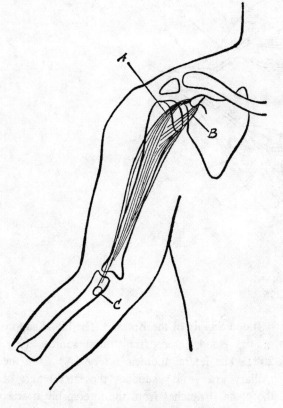

Fig. 203

three-headed muscle is found (Fig. 204). In this posterior view the long head arises from the scapula just below the glenoid fossa (A). The lateral head (B) is near the lateral side of the humerus, and the medial head (C) is near the medial side of the humerus. The three heads unite to form the **triceps muscle,** which is inserted by a single tendon onto the **olecranon process** (D) of the **ulna.** The olecranon process is the most prominent bony point on the posterior aspect of the elbow. As seen from a side view (Fig. 205), the muscle when contracting would straighten or extend the forearm.

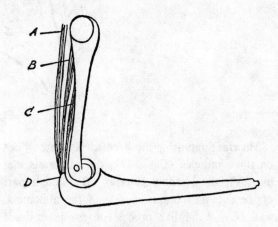

Fig. 205

THE BREAST

The female breast (Fig. 206) is made up of fifteen or twenty lobes of glandular tissue (D) drained by **lactiferous ducts** (E). The ducts enlarge (F) to form the **lactiferous sinuses,** where milk can be stored, just before they open on the nipple (G). The lobes are surrounded by fibrous tissue that supports the glandular tissue. Wedged between the lobes are collections of fat (H) that give the female contour to the breast. Strands of fibrous tissue continuous with that surrounding the lobes extend to the skin to be attached there (I). These are called the **suspensory ligaments of Cooper.** Cancer of the glandular tissue involves the fibrous tissue, producing tension on the suspensory ligaments and causing small dimples in the skin resembling the skin of an orange. This dimpling is usually a late sign of cancer of the breast. The breast is developed in superficial fascia, and this is separated from the underlying muscle (A) by the deep fascia of that muscle (B) as well as by the deep layer of superficial fascia (C). The normal breast is freely movable over the contracted muscle. In cancer of the breast the muscle and fascia are frequently involved in the cancerous process, fixing the breast to the muscle and its fascia. Thus, another late sign of breast cancer is fixation of the breast to the chest wall.

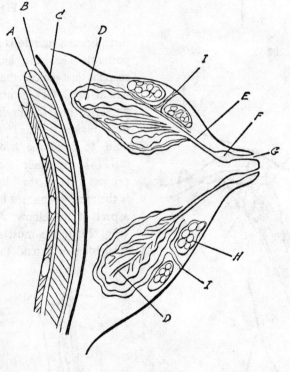

Fig. 206

Muscles Supporting the Breast. The breast lies on three muscles (Fig. 207), the **pectoralis major** (A), the **serratus anterior** (B), and a part of the external oblique muscle of the abdominal wall (C). A tail-like process of glandular tissue (D) almost always extends into the axilla.

Blood Supply of the Breast. The blood supply of the gland comes from three sources (Fig. 208). The lateral thoracic artery (A) from the axillary artery (B) supplies the lateral part of the gland. Branches from the intercostal arteries (C) pierce the chest muscles to enter the deep

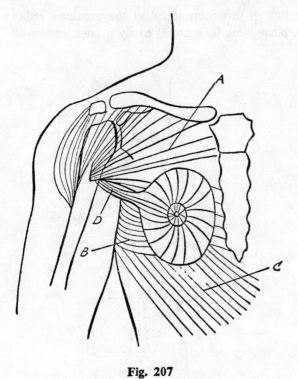

Fig. 207

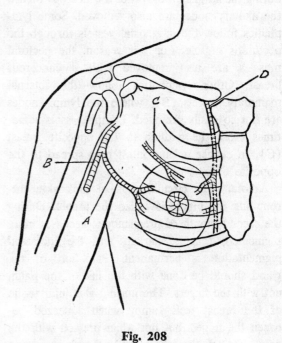

Fig. 208

surface of the gland, while branches from the **internal mammary artery** (D) enter the medial part of the breast.

The venous drainage follows the arterial supply and indicates the routes of the lymphatic drainage of the breast (Fig. 209). The major

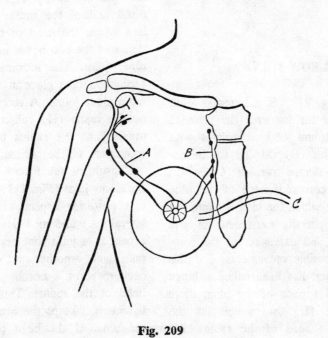

Fig. 209

lymphatic drainage is to the axilla (A). Thus, during the surgical removal of a cancerous breast, the axillary nodes are also removed. Some lymphatics follow the intercostal vessels through the pectoralis muscle. For this reason, the pectoral muscles are also removed with a cancerous breast. Other lymphatics follow the internal mammary vessels (B), where the lymph nodes are not so easily removed. Lymph vessels sometimes cross the midline to the opposite breast (C), so cancer in one breast may spread to the opposite one.

Areola. The circular area of pink skin surrounding the nipple is called the **areola.** During the second month of pregnancy it becomes more pigmented, that is, darker, and the increased pigmentation is permanent. Palpation of the gland should be done with the flat of the palm, not with the fingers. The normal glandular tissue of the breast feels lumpy when squeezed between the fingers but not when pressed with the palm against the chest wall. Regular, routine self-examination of the breast is the best method for the early detection of cancer.

THE ELBOW JOINT

The elbow joint (Fig. 210) is a synovial joint. At the distal end of the humerus the rounded surface of the **capitulum** (A) articulates with the head of the radius (B). Medial to the capitulum is the pulley-shaped **trochlea** (C), which articulates with the coronoid notch of the ulna (D). The manner by which the ulna fits into the V of the trochlea permits movement in one plane alone; flexion and extension of the forearm are the only possible movements. For this reason, the elbow joint has been called a **hinge joint.** On the lateral surface of the ulna, there is a rounded notch (E) into which fits the rounded edge of the head of the radius. Al-

though this joint is called the **proximal radioulnar joint,** its synovial cavity is continuous with

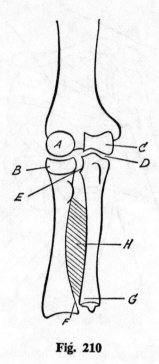

Fig. 210

the synovial cavity of the elbow joint. At the distal end of the radius there is a notch (F) into which the head of the ulna (G) fits. This union of the two bones is called the **distal radioulnar joint.** The movements of pronation and supination take place at the proximal and distal radio-ulnar joints. A tough sheet of fibrous connective tissue (H) called the **interosseous membrane** joins the radius to the ulna, and limits movement of the radius.

Capsule of the Elbow Joint. The capsule of the elbow joint (Fig. 211) is thin on both anterior and posterior aspects, but thickened on both lateral (A) and medial (B) sides, as you might expect in a joint that permits movement in only one plane. Another part of the capsule is thickened to form a fibrous noose (C) around the head of the radius. This is called the **annular ligament;** it keeps the head of the radius against the ulna. If the head of the radius were re-

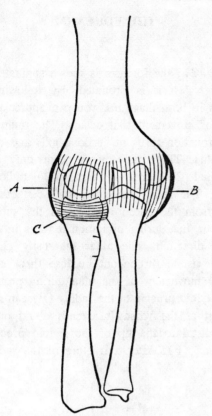

Fig. 211

of the radius is too small for the size of the annular noose. When a child is picked up by his hands or forearms there is a danger of pulling the head of the radius out of the annular ligament, producing a dislocation.

Fractures and Dislocations of the Elbow. The relative positions of the bony prominences at the elbow are often important in determining whether a fracture or dislocation has occurred. Extending the forearm (Fig. 213) and feeling the posterior aspect of the joint, you will find three bony points in the same transverse line. The prominent medial epicondyle (A), the olecranon, or bony point, of the elbow (B), and the lateral epicondyle (C) lie in a straight line (D). When

moved, the annular ligament would appear as shown in Fig. 212. In young children the head

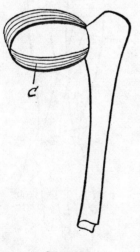

Fig. 212

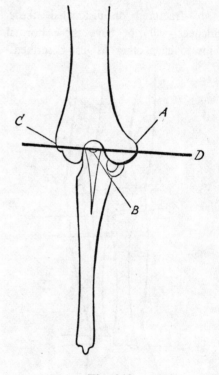

Fig. 213

the forearm is flexed (Fig. 214), the medial epicondyle (A), the olecranon (B), and the lateral epicondyle (C) form the apexes of an equilateral triangle. If the bones at the elbow

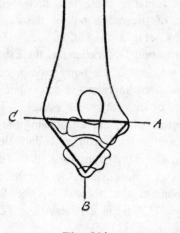

Fig. 214

THE FOREARM

In Fig. 215 the forearm is shown supinated and in Fig. 216 it is pronated. By following the radius in both diagrams, you can appreciate the type of movement that occurs. The radius during pronation turns on its long axis and crosses the ulna. These movements occur only at the proximal (A) and the distal (B) radio-ulnar joints. The radial tuberosity (C) can easily be seen from the anterior aspect in the supinated forearm, but during pronation it turns in a posterior direction. The **ulnar tuberosity** (D) remains visible throughout because there is little if any movement of the ulna during pronation. The **styloid process** of the radius (E) can readily be felt at the wrist. It extends about one-half inch distal to the tip of the styloid process of the ulna (F). After a fracture of the radius at

joint are fractured or dislocated, these bony prominences will not have these normal relationships to each other as just described.

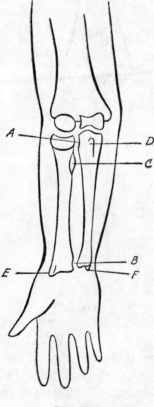

Fig. 215

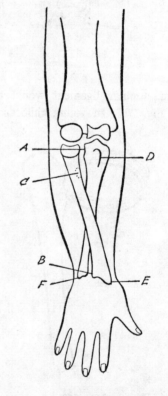

Fig. 216

the wrist, the styloid process of the radius may not extend as far distal as that of the ulna, because a fractured bone is frequently shorter. A very common type of fracture at the wrist is a **Colles' fracture** of the radius (Fig. 217). It

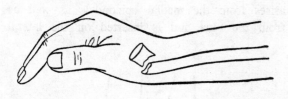

Fig. 217

usually results from a fall on the outstretched hand that pushes the distal fragment in a posterior direction. The shape of the wrist immediately after a Colles' fracture is very characteristic and has been called the **silver fork deformity.** Its resemblance to the silhouette of a silver fork is seen in Fig. 218.

Fig. 218

Muscles of the Forearm. Two muscles that flex the forearm are shown in Fig. 219. The **brachialis muscle** (A) arises from the anterior surface of the humerus and is attached to the ulnar tuberosity (B). The more superficial muscle, the biceps (C), is inserted into the radial tuberosity and is a powerful flexor of the forearm. The radius is not shown in this diagram.

Two muscles supinate the forearm. In Fig. 220 the forearm is pronated. The supinator mus-

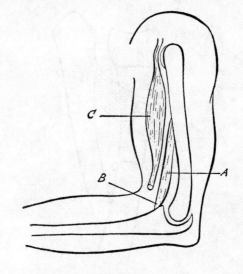

Fig. 219

cle (A) arises from the ulna and is wrapped around the lateral side of the radius. The en-

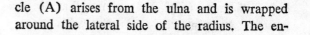

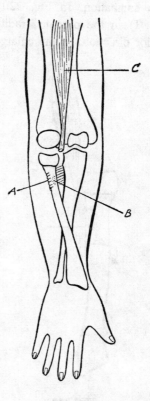

Fig. 220

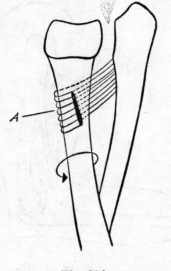

Fig. 221

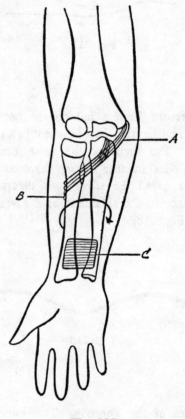

(Fig. 222) the tendon of the biceps (C) is likewise pulled in a posterior direction by its tubercle of insertion (B). When the biceps contracts, it not only rotates the radius in the direction indicated by the arrow (supination), but it also can flex the forearm.

In Fig. 223 two muscles that produce pronation are shown. The **pronator teres muscle** (A) arises from the medial epicondyle as well as from the ulna and is inserted on the lateral

larged diagram (Fig. 221) shows that if the muscle (A) contracts, the radius will be rotated in the direction indicated by the arrow, resulting in supination. In Fig. 220 the radial tuberosity (B) in the pronated radius is facing in a posterior direction. In the enlarged diagram

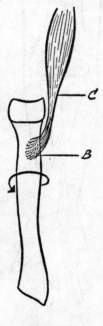

Fig. 222

Fig. 223

border of the radius (B). The other muscle, the **pronator quadratus** (C), arises from the ulna and inserts into the radius. Contraction of these muscles will rotate the radius in the direction indicated by the arrow, resulting in pronation. The superficial layer of muscles on the an-

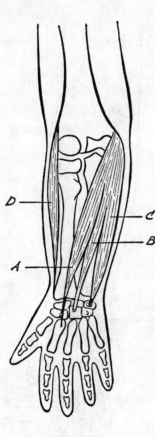

Fig. 224

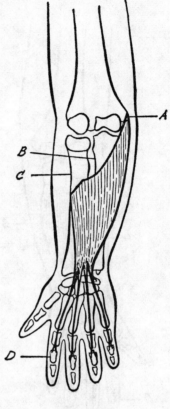

Fig. 225

terior aspect of the forearm (Fig. 224) arises from the medial epicondyle of the humerus. The **flexor carpi radialis** (A) is inserted into the base of the second metacarpal. The **palmaris longus** (B) is frequently absent. Its tendon, cut in this diagram, is attached to the **palmar aponeurosis** of the hand. The medial muscle is the **flexor carpi ulnaris** (C), which inserts into the **pisiform bone.** All three muscles flex the hand, and the flexor carpi ulnaris also adducts the hand. Along the lateral border of the forearm the **brachio-radialis muscle** (D) arises from the lateral epicondyle and inserts onto the distal end of the radius. It is a flexor of the forearm.

In Fig. 225 the superficial layer of muscles has been removed to show the middle layer. It consists of a single broad muscle, the **flexor digitorum sublimus,** arising from the medial epicon-

dyle (A), the ulna (B), and the radius (C). Four tendons emerge from its distal end and run to the fingers. Each tendon splits just before it inserts into the base of the **middle phalanx** (D). This muscle flexes the fingers; if the fingers are prevented from flexing, the muscle will flex the hand.

When the flexor digitorum sublimus muscle is removed, the third layer of muscles is exposed (Fig. 226). The lateral muscle (A) is the **flexor pollicis longus.** It arises from the radius and inserts into the base of the **distal phalanx** of the thumb (C). The **flexor digitorum profundus** (B) arises from the ulna and gives rise to four tendons. Each tendon inserts into the base of the **terminal phalanx** of a finger (D). This muscle aids the sublimus in flexion of the fingers.

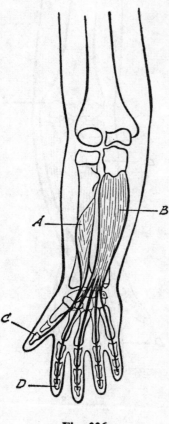

Fig. 226

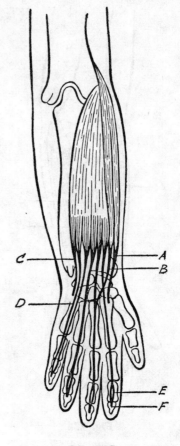

Fig. 227

The superficial muscles on the posterior aspect of the forearm are shown in Fig. 227. The **extensor carpi radialis longus** (A) and **brevis** (B) muscles pass distally to be inserted into the bases of the second and third metacarpals. Both muscles extend the hand. The **extensor carpi ulnaris** (C) inserts into the base of the fifth metacarpal. It not only extends but also adducts the hand. The main extensor of the fingers, the **extensor digitorum communis**, arises from the lateral epicondyle, and its four tendons are inserted into the **extensor expansion** (E), which inserts into the base of the middle as well as the distal phalanx (F). A second extensor tendon for the little finger (D) is frequently found.

When the superficial muscles are removed, the deep group is exposed (Fig. 228). The **abductor** **pollicis longus** (A) goes to the base of the first metacarpal. The **extensor pollicis brevis** (B) inserts into the base of the proximal phalanx. The **extensor pollicis longus** (C) passes to the base of the distal phalanx of the thumb. The **extensor indicis proprius** (D) provides a second extensor tendon to the index finger.

BRACHIAL PLEXUS

The nerves that supply the muscles of the shoulder girdle and upper limb arise in the neck. These nerves also contain fibers that carry sensations from the skin, muscle, bone, and joints

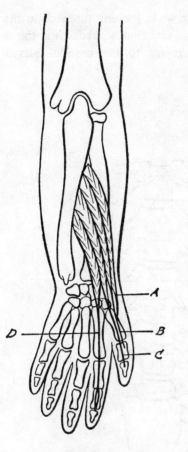

Fig. 228

spinal nerves are named from the vertebra just superior to their exit.) The fifth and sixth cervical roots combine to form the upper trunk (T). The middle trunk is a continuation of the seventh root, and the lower trunk is formed by the union of the eighth cervical and first thoracic roots. Each trunk now divides into an anterior and posterior division (D). The posterior divisions are shown cross-hatched in this diagram. The anterior divisions of the upper and middle trunks form the **lateral cord** (LC). The anterior division of the lower trunk forms the **medial cord** (MC), and the three posterior divisions form the **posterior cord** (PC).

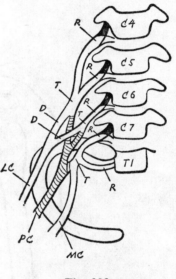

Fig. 229

back to the spinal cord. The nerves form a network, or plexus, by dividing and recombining (Fig. 229). Although on first sight the pattern appears complicated, it is fairly constant. This network, called the **brachial plexus,** begins as roots (R), which are formed by the ventral rami of the fifth, sixth, seventh, and eighth cervical nerves and the first thoracic nerve. The fifth cervical nerve emerges from the vertebral canal between the fourth (C4) and fifth (C5) cervical vertebrae. The eighth cervical nerve emerges between the seventh (C7) cervical vertebra and the first thoracic vertebra (T1). (In the cervical region a spinal nerve gets its numerical name from the vertebra immediately inferior to its exit from the spinal canal, with the exception of the eighth cervical nerve. All the remaining

The relationships of the brachial plexus to the bony structures in the lower neck and axilla are shown in Fig. 230. The fifth cervical root (A) combines with the sixth root to form the **upper trunk** (B). The anterior divisions of the upper and middle trunks form the lateral cord (C). The anterior division of the lower trunk forms the medial cord (E), and the posterior divisions form the posterior cord (D). The three cords pass from the neck to the axilla, posterior to the clavicle (F), but anterior to the first rib.

Roots or trunks are frequently torn by a type of injury that stretches the nerves. Carrying a heavy weight on the shoulder, depressing the shoulder while pushing the neck to the opposite side, or strenuously abducting the arm have caused injury to the brachial plexus.

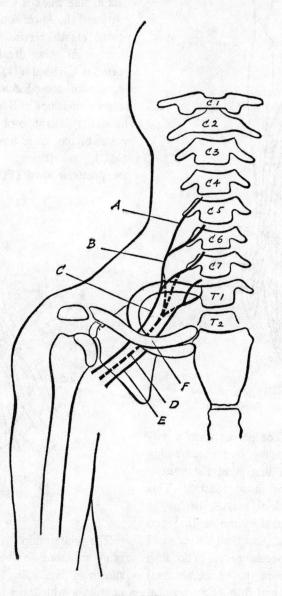

Fig. 230

NERVES SUPPLYING THE LIMB MUSCLES

All the muscles associated with the shoulder girdle are supplied by branches of the brachial plexus. The lateral (LC) and medial (MC) cords of the brachial plexus are shown in Fig. 231. The **median nerve** (C) arises from two heads (A and B) derived from the medial and lateral cords. A branch of this nerve (K) supplies all the flexor muscles of the forearm with the exception of the flexor carpi ulnaris and the medial half of the flexor digitorum profundus. Other branches of the median nerve supply the **thenar** muscles (L) and the skin (M) of the palm and the lateral 3½ fingers.

The **ulnar nerve** (D) is a major branch of the medial cord. It passes distally to lie posterior to the medial epicondyle (E), where it is quite superficial. At this point it can be felt by being rolled against the bone. Because of its vulnerable position, the nerve is frequently injured by trauma to the elbow.

part of the hand and 1½ fingers. The broken line marks the dividing line between those areas of skin supplied by the median and ulnar nerves. The major branch of the lateral cord is the **musculo-cutaneous nerve** (I), which supplies the biceps and brachialis and the skin on the lateral side of the forearm. Branches from both medial and lateral cords supply the pectoralis major muscle.

Axillary and Radial Nerves. One major branch of the posterior cord (Fig. 232) is the **axillary nerve** (A), which passes inferior to the

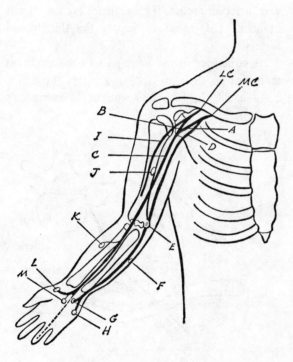

Fig. 231

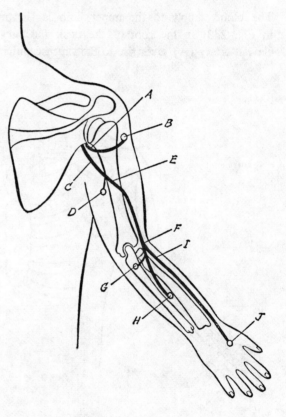

Fig. 232

The medial epicondyle has been called the "funny bone" because of the sensation produced by pressure on the ulnar nerve. In the forearm a branch of the ulnar nerve (F) supplies the 1½ muscles not supplied by the median nerve. The deep branch of the ulnar (G) supplies most of the small muscles of the hand. The superficial branch (H) supplies the skin over the medial

gleno-humeral joint and is injured in dislocations at that joint. The nerve winds around the posterior surface of the humerus to supply the deltoid muscle (B). The other major branch of the posterior cord is the **radial nerve** (C), which gives branches to the triceps muscle (D) and lies in a groove on the posterior surface of the humerus (E). Fracture of the humerus at E may injure the radial nerve. Passing distal to the elbow (F), the radial nerve divides into a superficial (I) and a deep branch. The deep branch supplies the supinator and superficial extensor muscles (G) as well as the deep muscles (H). The superficial branch passes distally to supply the skin on the dorsum of the hand (J).

BLOOD SUPPLY TO THE UPPER LIMB

The blood supply to the upper limb is shown in Fig. 233. In the root of the neck the **subclavian artery** (A) gives rise to the **suprascapular**

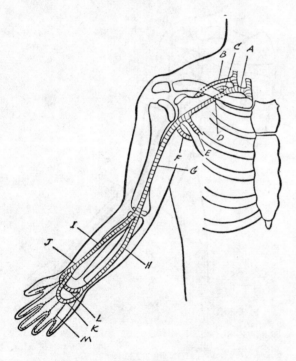

Fig. 233

(E) to the mammary gland and a **circumflex scapular artery** (F) that passes to the dorsum of the scapula. Here the axillary artery forms an important anastomosis with the suprascapular artery (B) and deep cervical artery (C). As the axillary artery enters the arm, the name is again changed, to the **brachial artery** (G). At the elbow the brachial artery divides into the **ulnar** (H) and **radial** (I) **arteries.** The radial artery at the point J is pressed against the radius by the fingers to "feel the pulse." The radial and ulnar arteries join each other in the hand to form the **superficial** (K) and **deep** (L) **palmar arterial arches.** From these arches digital branches (M) arise to supply the thumb and fingers.

As structures in the neck pass to the axilla on their way to the arm, they must pass through a bony canal. In Fig. 234 you are viewing this

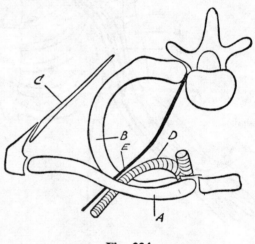

Fig. 234

artery (B) and the deep branch of the **transverse cervical artery** (C). Both descend on the posterior surface of the scapula. At the border of the first rib (D) the subclavian artery changes its name to the **axillary artery.** The axillary artery gives rise to a **lateral thoracic branch**

triangular-shaped canal from above, as if looking through the canal into the axilla. The clavicle (A) is the anterior bony boundary, the first rib (B) is medial, and the upper border of the scapula (C) is posterior. The subclavian artery (D) and its continuation, the axillary artery, are shown passing through the canal between the first rib and the clavicle, accompanied by the brachial plexus (E). Because the canal is shallow in this area, abnormalities of the clavicle or rib may interfere with the blood or nerve supply to the upper extremity. Traction on a person's arm, by pulling it downward and backward, may cause the clavicle to compress the artery against the rib with the result that no pulse can be felt in the radial artery. Bony enlargement at the site of a badly united fracture of the clavicle may also compress the artery. The first symptoms of compression are usually observed in the hand. Numbness, tingling, decrease in skin temperature, weakness, and ulcers that do not heal are suggestive of compression in the canal.

THE HAND

The eight carpal bones (Fig. 235) are arranged in two rows of four bones each. In the proximal row from lateral to medial side are the boat-

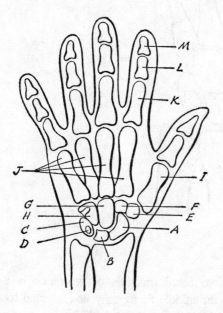

Fig. 235

shaped **navicular** (A), the half-moon-shaped **lunate** (B), and the pea-shaped **pisiform** (D) sitting on the **triquetrum** (C). In the distal row, the **trapezium** (E) articulates with the **first metacarpal** (I). The **trapezoid** (F), the massive **capitate**

(G), and the **hamate** (H) with its bony hook, articulate with the remaining **metacarpals** (J). A metacarpal consists of a **base** which articulates with a carpal bone, a narrow tapered **shaft,** and a **head** which articulates with a phalanx. Each finger has three phalanges: a **proximal** (K), a **middle** (L), and a **distal** (M). The thumb has only two phalanges, proximal and distal.

The wrist joint (A in Fig. 236) is formed

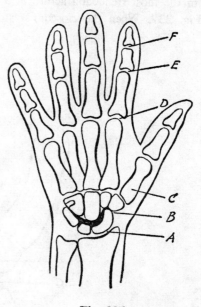

Fig. 236

between the proximal row of carpal bones and the radius. The ulna has little contact with the carpal bones. The transverse joint between the proximal and distal rows of carpal bones is called the **midcarpal joint** (B). Considerable movement occurs at this joint during flexion of the wrist; hence, disease of this joint seriously limits wrist flexion. The carpo-metacarpal joints allow for various degrees of movement. The first metacarpal bone (C) is the most mobile, while the third metacarpal is the least mobile. This is a possible reason for the more frequent injury of the third metacarpal.

The joint between a metacarpal and the proximal phalanx is called the **metacarpo-phalangeal joint** (D); this term is sometimes abbreviated to

MP joint. The joints between the proximal and middle phalanges are called the **proximal interphalangeal joints** (E) or **proximal IP joints,** and those between the middle and distal phalanges are called the **distal IP joints** (F).

The prominent knuckles observed when you make a fist are formed by the heads of the metacarpals. Examine the bony architecture of the hand and visualize the mechanism which results in the most frequent fracture of a carpal bone, Fig. 237. When the closed fist strikes a solid object, the major force (A) is taken by the most prominent knuckle, the head of the third metacarpal. The force is transmitted down the shaft of the metacarpal to the capitate (B). The narrow waist of the navicular (C), compressed between capitate and the radius, breaks into two parts.

Movements and Muscles of the Hand. To produce flexion of the fingers, make a fist. Flexion occurs not only at the IP joints but also at the MP joints. Extension or straightening of the fin-

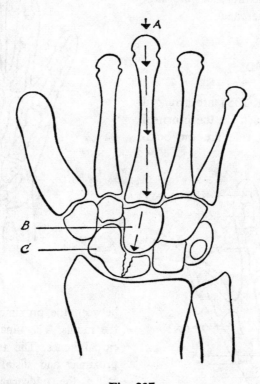

Fig. 237

gers is the opposite of flexion. Abduction and adduction of the fingers occur at the MP joints. The axis of reference for these movements is an arbitrary line passing distally along the middle or third finger. Spread your fingers to produce abduction of the index, ring, and little fingers, that is, movement away from the axis of reference—the middle finger. The opposite movement of bringing the fingers close together is called adduction. Since the axis of reference is a fixed line, the middle finger may be abducted to either side of this line: toward the index finger or toward the ring finger.

The movements of the thumb resemble those of the fingers, but they occur in a plane at right angles to the fingers. Place your thumb close to the palm of your hand with thumbnail at right angles to the fingernails. Abduction of the thumb

(Fig. 238) is the movement of the thumb anterior to the palm of the hand; adduction is the reverse movement. Flexion of the thumb

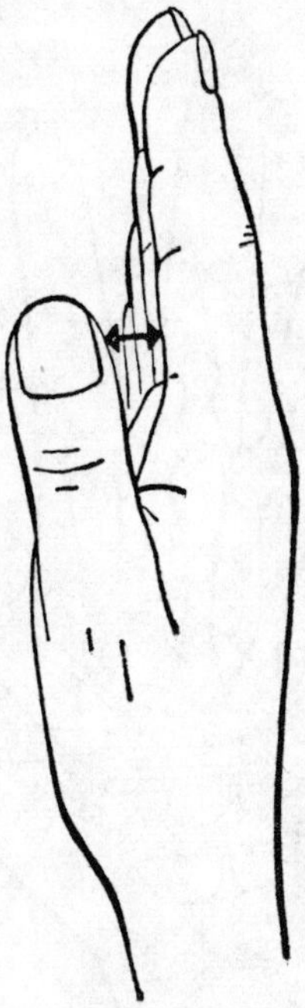

surface of the carpal bones they are held against the bones by a strong ligament called the **flexor**

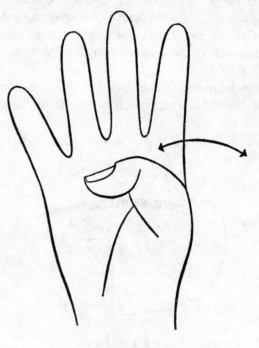

Fig. 239

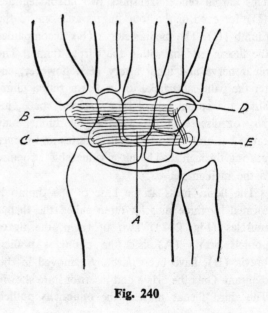

Fig. 238

(Fig. 239) is the bending of the thumb across the palm, keeping contact with the palm during the movement. Extension of the thumb is the reverse, to the full extent of the movement.

The most important movement of the thumb **is opposition.** This is produced by having the tip of the thumb touch the tip of any one of the fingers. Any pincer movement of the thumb and fingers, for example the holding of a pencil in writing, requires the movement of opposition of the thumb. At least 30 percent of the efficiency of the hand for manual labor is lost if one is unable to perform this movement. Opposition of the thumb has been described as the single most important movement of the hand.

The long flexor tendons of the fingers and thumb, described previously as arising from the ventral compartment of the forearm, cross the wrist to enter the palm. As they cross the ventral

retinaculum, shown in Fig. 240. The flexor retinaculum (A) prevents "bow stringing" of the

Fig. 240

flexor tendons during flexion of the fingers. The four corners of the retinaculum are attached to the carpal bones. On the lateral side they are attached to the trapezium (B) and navicular (C). On the medial side they are attached to the hook of the hamate (D) and the pisiform (E).

A transverse section through the wrist is shown in Fig. 241. The ventral surface of the carpal

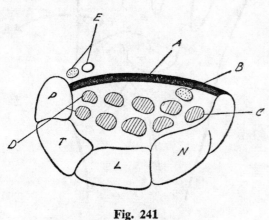

Fig. 241

bones, pisiform (P), triquetrum (T), lunate (L), and navicular (N), form a concavity which is roofed over by the flexor retinaculum (A). This carpal tunnel transmits two flexor tendons (D) for each of the four fingers and one to the thumb (C). The median nerve (B) accompanies the flexor tendons within the carpal tunnel. The ulnar nerve and ulnar artery (E), however, enter the palm superficial to the flexor retinaculum. Since the carpal tunnel is a confined space, injury or disease of the walls of the tunnel may cause harmful pressure on the arteries and nerves passing through it. The most vulnerable structure is the median nerve.

The fleshy mass at the base of the thumb is formed by three muscles often called the **thenar muscles** (Fig. 242). Two of them, the **flexor pollicis brevis** (A) and the **abductor pollicis brevis** (B), have been partially removed in this diagram. Only the origin and insertions are shown. The third thenar muscle, the **opponens pollicis**

(C), is inserted into the shaft of the first metacarpal, while the other two insert onto the base of the proximal phalanx. All three arise from the lateral edge of the flexor retinaculum. A

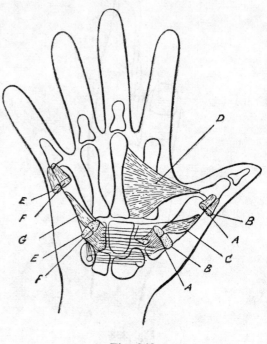

Fig. 242

deeper muscle in the palm, which cannot be readily palpated, also inserts on the proximal phalanx of the thumb. It is the **adductor pollicis** (D), which arises from the middle metacarpal. The smaller fleshy mass at the base of the little finger is formed by the **hypothenar muscles,** which have the same relation to the little finger that the thenar muscles have to the thumb. The **abductor digiti minimi** (E) and **flexor digiti minimi brevis** (F) have been partially cut away to show the **opponens digiti minimi** (G) inserting on the shaft of the fifth metacarpal. The names of these muscles indicate their functions. You can appreciate that the movements produced by the hypothenar muscles are not of the same functional importance as those produced by the thenar muscles.

The deepest muscles of the palm are the interossei, shown in Fig. 243, which fill the spaces between the metacarpals. The three **palmar interossei** (P) send their tendons to the dorsum of the fingers. The four **dorsal interossei muscles** (D) lie immediately posterior to the palmar interossei. With the exception of the first dorsal interosseous, they insert on the base of the proximal phalanges as well as onto the dorsum of the fingers. A useful guide for remembering the

joint, and because of its relation to the head of the metacarpal is called the **hood** of the extensor expansion. The hood receives the interossei muscles (C) and the **lumbrical muscle** (D), as well as the long extensor tendon (E). The remaining part of the extensor expansion separates into three bands. The central band inserts on the base of the middle phalanx (F) while the outer bands insert on the base of the distal phalanx (G).

The interossei and lumbrical muscles, by pulling on the sides of the hood, flex the MP joint;

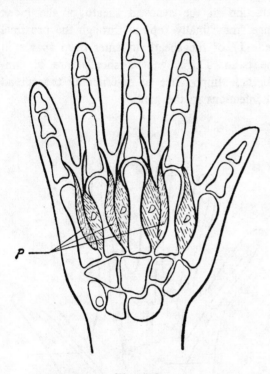

Fig. 243

functions of the interossei muscles is found in the terms **"pad"** and **"dab."** That is, the palmar interossei muscles **ad**duct the fingers **("pad"),** and the **d**orsal interossei muscles **ab**duct the fingers **("dab")**. The interossei also flex the MP joints and extend the IP joints by means of the mechanism shown in Fig. 244. The wide fibrous band on the dorsal surface of the phalanges is called the **extensor expansion** (A), that is, the expanded part of the extensor tendon of the finger. The broad proximal part (B) is draped over the dorsal surface and sides of the MP

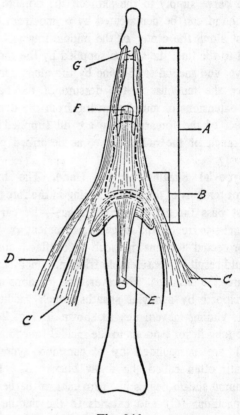

Fig. 244

by exerting tension on the extensor expansion they can extend the IP joints. The long extensor of the finger (F) extends the MP joint by pulling on the hood, or it can exert tension on the expansion to extend the IP joints. There is a rather complex interplay between the long extensor tendon and the interossei muscles during extension of the IP joints.

Nerves of the Hand. As we saw in Fig. 241, two major nerves enter the palm. The median nerve supplies the major part of the thenar muscles, while the ulnar nerve supplies the remaining intrinsic muscles of the hand. If a line is drawn from the wrist distally on the palm and along the middle of the ring finger, the skin medial to the line will be supplied by the ulnar nerve, and lateral to the line by the median nerve. The skin over the dorsum of the middle and distal phalanges is also supplied by these palmar nerves. The nerve supply to the skin on the dorsum of the hand can be demarcated by a line from the wrist along the center of the middle finger. Lateral to the line, the skin is supplied by the radial nerve, and medial to the line by the ulnar nerve. Since the tendons on the dorsum of the hand are extensions of muscles arising from the dorsal aspect of the forearm, they are all supplied by a branch of the radial nerve, as described previously.

Synovial Sheaths of the Hand. The long flexor tendons of the fingers arising in the forearm must pass through a narrow space—the carpal tunnel—to reach the fingers. Movement of the unprotected tendons in such a confined space would result in considerable friction, which could injure the tendons. However, the tendons are enveloped by synovial sheaths which facilitate their gliding movements, as shown in Fig. 245. The long flexor tendons to the medial four fingers (A) are surrounded by a common synovial sheath often called the **ulnar bursa** (B). The common sheath begins just proximal to the flexor retinaculum (C) and extends to the middle of the palm. Since the flexor tendons are also confined as they travel distally along the fingers, they are also protected by separate synovial sheaths (D). While the synovial sheaths of the index, middle, and ring fingers begin at the base of the finger, that of the little finger (E) is continuous with the ulnar bursa. The long flexor of the thumb (F) is likewise enclosed by a

separate synovial sheath called the **radial bursa** (G). The radial and ulnar bursae usually communicate with each other (H) deep to the flexor retinaculum. Synovial sheaths in the fingers may become infected from a puncture wound (a pinprick, for example), and the infection readily spreads along the synovial sheath. Infection of the synovial sheath of the little finger may spread directly to the ulnar bursa and then through the synovial communication (H) to the radial bursa. Infection in the synovial sheath of the index finger may finally rupture through the proximal end (I) of the sheath to enter deep spaces in the palm. Thus, synovial sheaths are of considerable importance in determining the spread of infections in the hand.

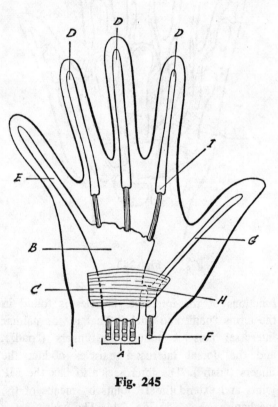

Fig. 245

Protective Coverings of the Flexor Tendons. Dense fascia covers the superficial aspect of the flexor tendons in the palm (Fig. 246) to protect them from the wear and tear to which the hand is exposed. Although this fascial covering is one continuous layer, it is thicker in some areas and therefore given a special name. The central part of the palm is covered by dense fascia called the **palmar aponeurosis** (A). It extends from the flexor retinaculum (B), where it receives the palmaris longus tendon (C), to the bases of the fingers, where it divides into two slips which are attached to the fibrous sheaths of the fingers (D).

Less dense fascia covers the thenar muscle (E) as well as the hypothenar muscles (F). The fibrous sheaths of the fingers hold the flexor tendons against the phalanges. The cut ends of the two flexor tendons for the ring finger (G) may be seen entering its fibrous sheath. The fibrous sheath of the finger does not have a uniform thickness. It is very thin where it overlies an IP joint (H) and thicker between the finger joints (I) so as not to hinder joint movement. The MP and IP joints of the middle finger are outlined in relation to the thin and thick portions of the fibrous sheath.

A cross section of a finger is shown in Fig. 247 to illustrate the relation of the synovial

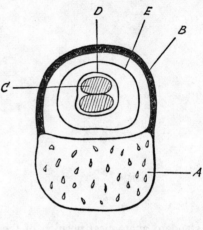

Fig. 247

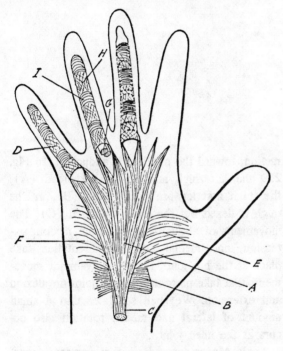

Fig. 246

sheath to the fibrous sheath of a flexor tendon. The bony phalanx (A) forms the floor of a tunnel for the tendon. The roof is formed by the fibrous sheath (B) which is attached to the edges of the phalanx. Within the tunnel the two flexor tendons (C) are closely surrounded by a layer of synovial membrane (D), while the tunnel is lined by another layer of synovial membrane (E). The two layers, D and E, constitute the synovial sheath.

CHAPTER 9

THE LOWER EXTREMITY

In the standard anatomical position of the lower limb (Fig. 248), the limbs and feet are together and the toes point forward. The **thigh** (A) extends from the groin to the knee, and the **leg** (B) from the knee to the ankle. Distal to the ankle is the **foot** (C). The area of the **hip joint** (D), the **knee joint** (E), and the **ankle joint** (F) complete the lower segment.

MOVEMENTS OF THE LOWER EXTREMITY

Thigh Movements. Medial rotation of the thigh is performed by turning the lateral side of the thigh toward the midline (arrow G in Fig. 248). Lateral rotation is the reverse movement (arrow H). Abduction of the thigh is shown in Fig. 249. Abduction of the right thigh is movement away from the midline. Movement in the opposite di-

rection, toward the midline, is adduction. In Fig. 250 the left thigh and leg are extended (A), the thigh and leg are both flexed (B), or the thigh is flexed but the leg is extended (C). The movements of abduction, adduction, flexion, extension, and medial and lateral rotation take place at the hip joint, but the only major movements that take place at the knee joint are flexion and extension. We shall see later that a small amount of lateral and medial rotation also occurs at the knee joint.

Ankle Movements. Before movements occurring at the ankle joint (Fig. 251) are discussed, the surfaces of the foot should be identified. The superior aspect of the foot (A) is referred to as the **dorsum** of the foot, and the inferior surface (B) is called the **plantar aspect** of the foot. At C the foot is in the neutral position. Bending the dorsum of the foot closer to the front of the leg (D) is called **dorsiflexion,** literally, "flexion of the dorsum of the foot." The opposite movement (E) is called **plantar flexion** of the foot. No

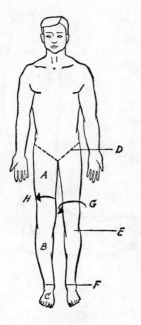

Fig. 248

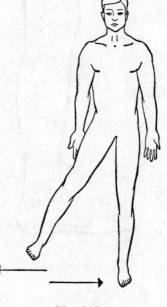

Fig. 249

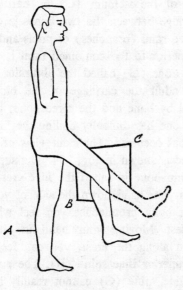

Fig. 250

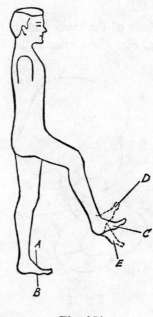

Fig. 251

other movements of the foot are possible at the ankle joint. However, the foot *can* perform those movements shown in Fig. 252. The neutral position is shown at A. At B, the plantar surface is turned outward. This is called **eversion** of the foot, literally, "a turning out." At C, the plantar surface is turned inward, called **inversion** of the foot. These movements of eversion and inversion take place not at the ankle joint but at the joints between the foot bones.

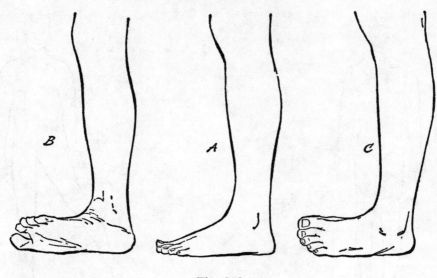

Fig. 252

PELVIC GIRDLE

The bony girdle of the lower limb is formed by the hip bone. Fig. 253 illustrates the external or

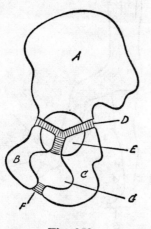

Fig. 253

lateral aspect of a child's hip bone. It consists of three bones joined by cartilage: the **ilium** (A), **pubis** (B), and **ischium** (C) are held together by the Y-shaped cartilage (D), which passes through the **acetabulum** (E). The acetabulum is the bony cup that receives the ball-shaped head

of the **femur**, or thigh bone. A barlike extension of the pubis (B) is also joined to a similar extension of the ischium (C) by cartilage (F). This bridge between the two bones is called the conjoined rami (branches) of pubis and ischium. Just superior to the conjoined rami is a hole in the hip bone (G) called the **obturator foramen.** In the adult the cartilage of the hip bone is replaced by bone and the three bones are fused, so that one has difficulty finding the lines where fusion has occurred. The same view of the adult hip bone is shown in Fig. 254. The superior rim of the hip bone is called the **iliac crest** (A). If you run your hand down the side of your trunk, the first bony prominence you feel will be the iliac crest. Moving your hand in an anterior direction along the crest, you can feel the **anterior superior iliac spine** (B). The **anterior inferior iliac spine** (C) cannot readily be felt in the living person. The conjoined rami (E) have fused, and immediately superior to them is the obturator foramen (F). No evidence of the cartilage seen in the child's acetabulum remains in the adult acetabulum (D). The bony prominence of the ischium (G), called the **ischial tuberosity,** is what we actually sit on. Posterior to the ischial tuberosity, a bony projection (I) called the **ischial spine** separates two indentations in the bone. The superior and larger one (J) is called the **greater sciatic notch;** the other (H) is called the **lesser sciatic notch.** The ilium, besides having two anterior bony projections (B and C), has two posterior projections: the **posterior superior iliac spine** (L) and the **posterior inferior iliac spine** (K). These spines mark the site of the sacroiliac joint.

sacroiliac joint. The sacroiliac joint is partly a synovial and partly a fibrous joint. The synovial part (J) is smooth, but the roughened area (K) is joined to the sacrum by tough fibrous connective tissue.

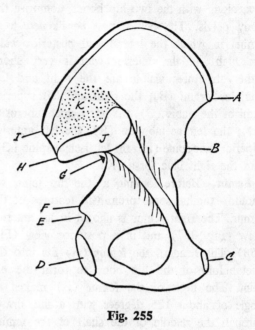

Fig. 255

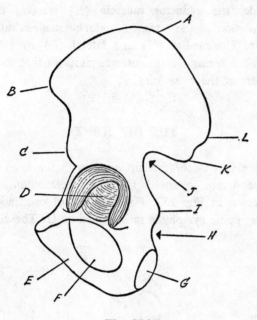

Fig. 254

The inner side or medial surface of the hip bone (Fig. 255) has several features that cannot be seen from the lateral aspect. Orient yourself by identifying the anterior superior iliac spine (A) and the inferior spine (B). The area C on the pubic bone is joined to the opposite pubic bone by fibrocartilage, forming the pubic symphysis. The ischial tuberosity (D), the lesser sciatic notch (E), the ischial spine (F), the greater sciatic notch (G), and the posterior inferior (H) and posterior superior (I) iliac spines have all been identified previously. The areas J and K articulate with the sacrum to form the

In Fig. 256 the right and left hip bones, seen in an anterior view, are in their normal position. The two pubic bones are held together by fibro-

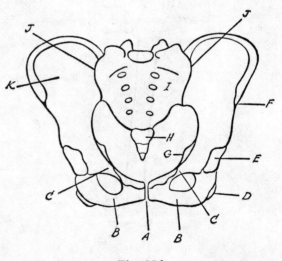

Fig. 256

cartilage (A) called the pubic symphysis. The two iliac bones unite with the sacrum (I) at the right and left sacroiliac joints (J). The several bony segments (H) at the inferior end of the sacrum form the **coccyx.** The sacrum and coccyx, along with the two hip bones, comprise the bony pelvis. The pelvis may be likened to a funnel in which the lateral and posterior walls are high but the anterior wall is very short. Other structures visible are the right and left conjoined rami (B), the right and left superior rami of the pubis (C), the left ischial tuberosity (D), the left acetabulum (E), the left anterior superior iliac spine (F), the left ischial spine (G), and the right iliac fossa (K).

Femur. Before looking at the hip joint, we should examine some prominent features of the femur. The right femur is shown in an anterior view (Fig. 257) and in a posterior view (Fig. 258). The head of the femur (A) fits into the acetabulum of the hip bone to form the hip joint. The neck of the femur (B) makes an angle of about 125 degrees with a line drawn through the middle of the shaft of the femur.

The enlargement (C) at the upper end of the shaft is called the **greater trochanter;** the smaller bony prominence (D) is called the **lesser trochanter.** A slightly raised bony ridge on the anterior aspect (H) runs from the greater to the lesser trochanter. This is called the **intertrochanteric line;** it indicates the attachment of the capsule of the hip joint. On the posterior aspect of the femur, a well-defined ridge (I) joins the trochanters and is called the **intertrochanteric crest.** At the distal end of the shaft on the medial side, the **adductor tubercle** (E) receives the insertion of the powerful **adductor magnus muscle.** The **medial** (G) and **lateral** (F) **condyles** of the femur are the articular surfaces that form part of the knee joint.

THE HIP JOINT

A side view of the hip bone, with the head of the femur removed from the acetabulum, is shown in Fig. 259. For purposes of orientation the pubic symphysis is indicated at H. The ace-

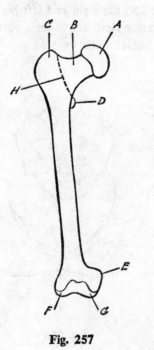

Fig. 257

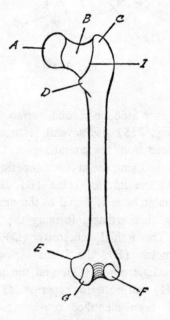

Fig. 258

tabulum (A) is a rather deep bony cup with a gap in the inferior part of the bony rim. This gap is filled by a band of fibrous connective tissue (B) called the **transverse acetabular ligament.** From the inner edge of this ligament there arises a short round ligament (C) called the **ligamentum teres.** (*Teres* means "round.") The ligament, although cut in this diagram, attaches to the head of the femur (F); the cut end of the ligament is shown at E. A small artery (D) arises from the obturator artery and runs in the ligamentum teres to supply a part of the bony head of the femur. Other arteries (G), arising from the femoral circumflex arteries, run along

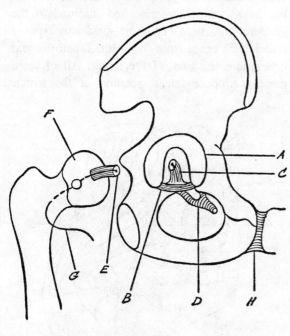

Fig. 259

the neck of the femur and enter the bone to supply both the femoral neck and femoral head. Later we will see how fractures of the femoral neck or dislocatons of the hip joint may cut off this blood supply, resulting in necrosis or death of the bony head of the femur.

Since the hip joint must bear the weight of the body during walking or standing, we must look at those structures of the joint that give it strength and stability. Contrast this joint with the gleno-humeral joint. Pliable structures surrounding the gleno-humeral joint insure its mobility. However, it has to pay the price for its lack of stability: the relative ease of shoulder dislocation. The bony acetabulum of the hip joint receives the major part of the head of the femur, thereby conferring a great deal of stability at the expense of mobility. In addition, the capsule of the hip joint is thickened in various parts to form strong ligaments.

On the anterior aspect of the joint (Fig. 260) the capsule is thickened to form the **ilio-femoral,** or Y-shaped, **ligament;** the two arms of the Y are shown at A and B. This ligament, the strongest ligament of the body, arises from the anterior inferior iliac spine (C) and attaches to the inter-trochanteric line. Another thickening of the cap-

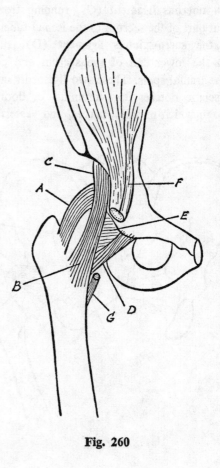

Fig. 260

sule is the **pubo-femoral ligament** (D), attached to the pubic bone as well as the femur. Between these two ligaments the capsule is relatively thin (E), but this area is covered by the tendon of the **iliopsoas muscle** (F), cut in this diagram. The **iliopsoas tendon** inserts into the lesser trochanter (G). You can appreciate that the deep bony cup plus the strong ligaments and the iliopsoas tendon confer great strength and stability to the joint. On the posterior aspect of the joint (Fig. 261) the capsule is thickened to form the **ischio-femoral ligament** (A), but it leaves uncovered the lateral part of the femoral neck (B).

Gluteal Region. Several muscles acting at the hip joint are found in the **buttock,** or gluteal region, as it is sometimes called. The bony skeleton of the gluteal region is shown in Fig. 261. Two ligaments complete the framework: the **sacro-tuberous ligament** (C), running from the dorsal part of the sacrum to the ischial tuberosity, and the **sacro-spinous ligament** (D), running from the lower end of the sacrum and coccyx to the ischial spine. These two ligaments convert the sciatic notches into foramina, or doorways, through which muscles, nerves, and vessels pass.

Note the greater sciatic foramen (E) and the lesser sciatic foramen (F).

Muscles of Gluteal Region. The key muscle in the gluteal region (Fig. 262) is the **piriformis** (A), which arises from the anterior surface of the sacrum, passes through the greater sciatic foramen, and is inserted into the greater trochanter. Appearing at its superior border are the **superior gluteal vessels** (B) and nerve. At its inferior border the large sciatic nerve (C) emerges and passes inferior to the piriformis into the thigh. Medial to the sciatic nerve are the **inferior gluteal vessels** (D) and nerve. The **obturator internus muscle** (E) arises from the pelvic surface of the pubic bone, emerges through the lesser sciatic foramen, and inserts into the greater trochanter. The short **quadratus femoris muscle** (F) arises from the ischial tuberosity and inserts into the back of the femur. All of these muscles produce lateral rotation of the femur.

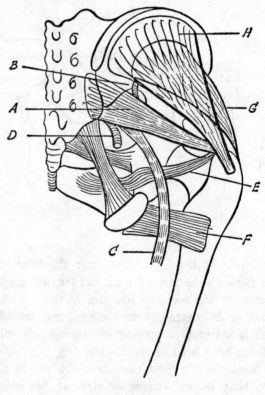

Fig. 262

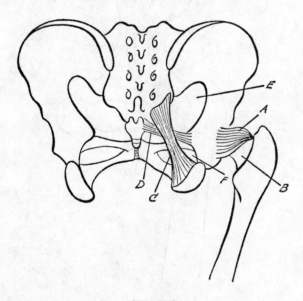

Fig. 261

Two muscles arise from the lateral surface of the hip bone and insert on the greater trochanter. The **deeper gluteus minimus** (G) and the more superficial **gluteus medius** (H) produce abduction of the thigh. The most massive muscle of the buttock (Fig. 263) is the **gluteus maximus** (A), which arises from both the ilium and the sacrum. Most of the muscle fibers insert into the **ilio-tibial tract** (B), which is the thickened lateral part of the deep fascial covering of the thigh muscles. A few of the deeper muscle fibers are inserted into the femur (C). The gluteus maximus muscle extends the thigh and also rotates the thigh laterally. A part of the gluteus medius muscle (D) can be seen before it passes deep to the gluteus maximus muscle.

Nerve Supply to Muscles. The nerve supply to all the muscles described in the gluteal region is derived from the **sacral plexus**. A person with one paralyzed gluteus maximus muscle cannot walk up a stairs or up an inclined plane. The gait of such a person is characteristic. When he bears his weight on the paralyzed limb, his body and arms are suddenly pulled back. He then makes a hurried step with the healthy limb so that the weight can quickly be taken off the affected limb. If both glutei maximus muscles are paralyzed, he cannot rise from a sitting position. He walks with his trunk thrust back to keep his center of gravity behind the hip joint. When his center of gravity falls in front of the hip joint, he is unable to prevent his trunk from bending forward at the hips.

An important function of the glutei medius

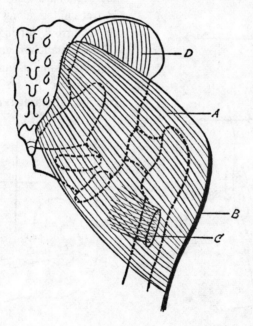

Fig. 263

and minimus muscles is shown in Figs. 264 and 265. The anterior superior iliac spines of a person standing on level ground (Fig. 264) would lie in the same tranverse plane (A). Similarly, both knee joints would lie in the horizontal plane (B). Just before the right limb takes a step the foot must clear the ground (Fig. 265). The left glutei medius and minimus muscles (C) contract to tilt the pelvis to the left. This tilting of the pelvis enables the right foot to clear the ground so that a forward step can be taken. Neither the anterior superior iliac spines (A) nor the

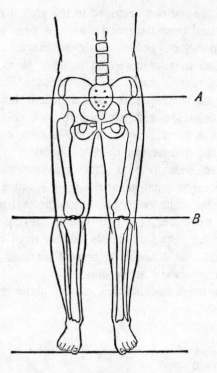

Fig. 264

knee joints (B) now lie in a horizontal plane.

Perhaps the most powerful muscle that flexes the thigh is the **iliopsoas** (Fig. 265). It is made up of the **psoas major muscle** (D), which arises from the lumbar vertebrae (G), and the **iliacus muscle** (E), which arises from the iliac fossa. The tendons of the two muscles fuse to form one tendon that is inserted on the **lesser trochanter** (F).

Injuries to the Hip Joint. The most common type of dislocation of the hip joint is a posterior dislocation of the head of the femur. It is usually the result of violent force, and is most frequently caused by an automobile accident. Typically, the victim is sitting in the front seat with knees crossed, and at the moment of impact (Fig. 266) the body is thrown forward (A), the knee strikes the dashboard (B), and the resultant force (C) is transmitted along the femur in the direction of the arrow, forcing the femoral head (D) out of the acetabulum and often breaking off a piece of the acetabular rim. As the head of the femur is forced to lie on the posterior aspect of the ischium, the thigh must assume a position of

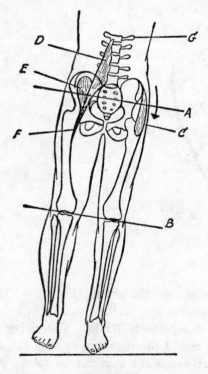

Fig. 265

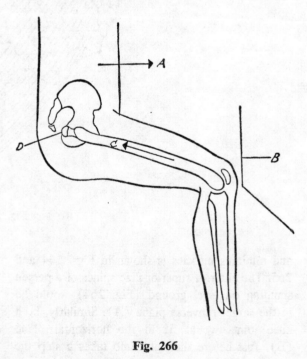

Fig. 266

adduction and medial rotation, because the very strong ilio-femoral ligament usually remains intact. An enlargement of this region (Fig. 267)

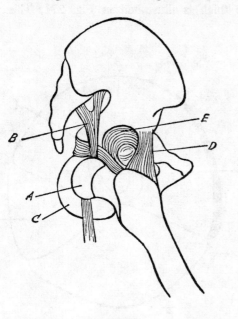

Fig. 267

shows how the head of the femur (A) may crush the sciatic nerve (B) against the ischium (C). The strong ilio-femoral ligament (D) is

still attached to the femur. The empty acetabulum is shown at E.

Congenital dislocation of the hip is more common in female than male infants. An infant's pelvis showing congenital dislocation of the right hip joint is seen in Fig. 268. The normal left hip joint shows the growing depression (A) in the bone that will later form the cuplike acetabulum. The head of the femur, with its secondary ossification center (B), fits into the depression. On the right side there is a very shallow depression (C) for the head of the femur. In addition, the ossification center (D) is deformed. As the infant attempts to stand or walk, the hip joint becomes dislocated. The deformed femoral head, in attempting to support the weight of the body, slides past the abnormally shallow acetabulum.

Fracture of the femoral neck is a frequent type of hip injury in the elderly. A fall or even a stubbing of the toe may fracture the neck of the femur. Bones of elderly people often lose calcium salts, with the result that they cannot withstand those stresses and strains that more youthful bones can endure. In earlier days the victim was usually confined to bed. As a result

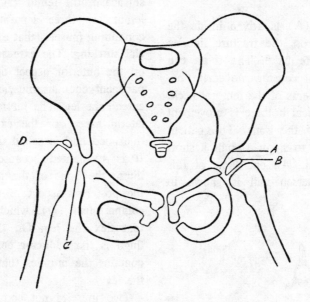

Fig. 268

of inactivity, death often occurred from lung infections or blood clots in the venous circulation. Today, early activity following the fracture is made possible by mechanically fixing the broken bone. The type of mechanical fixation most frequently used is the **Smith-Petersen nail** (Fig. 269). Through an incision in the lateral side

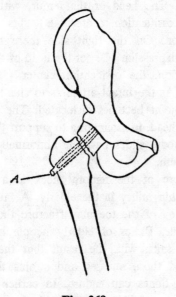

Fig. 269

of the thigh, the nail (A) is driven inside the neck of the femur across the fracture line so that both fragments are fixed. The thigh can then be moved without causing movement between the broken fragments of the femoral neck. The fracture will not heal if there is movement of the broken ends of the bone. The patient may even be permitted to sit in a chair a short time after the operation. An end view of this three-flanged Smith-Petersen nail is shown in Fig. 270.

Fig. 270

THE THIGH

Muscles. The general plan of the muscles of the thigh is illustrated in Fig. 271. This is a

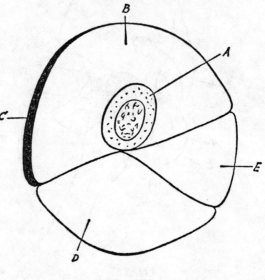

Fig. 271

cross section through the thigh showing the cut surface of the femur (A). Grouped around the femur are three separate compartments, each containing muscles that are surrounded by a **fascial stocking**. The **extensor compartment** (B) is on the anterior aspect of the thigh. This compartment contains muscles whose function is to extend the leg. The fascial stocking covering the lateral aspect of the extensor compartment is thickened and is called the **ilio-tibial tract** (C). It is so named because it extends from the ilium, the bone of the pelvis, to the **tibia**. On the posterior aspect of the thigh is the **flexor compartment** (D), which contains the muscles that flex the leg. On the medial side of the thigh is the **adductor compartment** (E), which contains the muscles that adduct the thigh, not the leg.

Two muscles not considered part of the extensor compartment are shown in Fig. 272, although they are on the anterior aspect of the thigh. The **sartorius muscle** (A) arises from the anterior superior iliac spine and inserts on the tibia (B). The **tensor fasciae latae** (C) arises from the ilium and inserts into the lateral fascia of the thigh (D). It was previously mentioned that this lateral fascia was thickened and called the ilio-tibial tract. Both muscles flex the thigh.

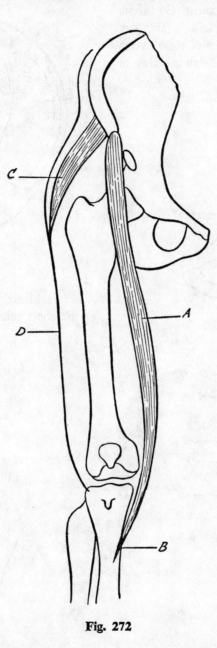

Fig. 272

The four muscles of the extensor compartment are shown in Fig. 273. The longest one, the **rectus femoris,** arises from the anterior inferior iliac spine (A) and is cut to show the underlying muscle. The other end of the muscle (B) inserts into the **patella** (F). The remaining three muscles are grouped around the femur. The **vastus lateralis** (C), the **vastus medialis** (D), and the **vastus intermedium** (E) all arise from the femur and insert into the patella. From the patella a ligament called the **patellar ligament** (G) inserts into the tibia. When these muscles contract, they extend the leg. The arrangement of the muscles as seen in a cross section of the thigh is shown in

femoris (F). The thickened ilio-tibial tract is indicated at D.

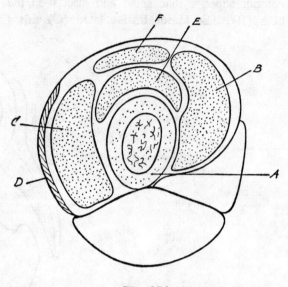

Fig. 274

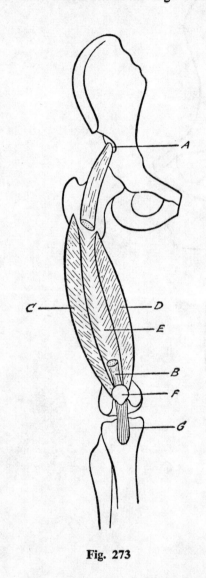

Fig. 273

Before identifying the adductor muscles, look at the short muscle (A) in Fig. 275; it arises

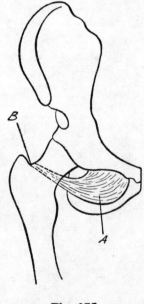

Fig. 275

Fig. 274. Grouped around the femur (A) are the vastus medialis (B), the vastus lateralis (C), the vastus intermedius (E), and the rectus

from the hip bone and passes posterior to the neck of the femur to insert into the greater trochanter (B). Since it arises in the area external to the obturator foramen, it is called the **obturator externus muscle**; it laterally rotates the thigh.

The adductors of the thigh are shown in Fig. 276. The **pectineus** (A) and **adductor longus** (C) muscles lie in the same plane. Just deep to the adductor longus is the **adductor brevis muscle** (B). The deepest muscle is the large **adductor magnus** (D). Its medial border inserts into the adductor tubercle of the femur. Where

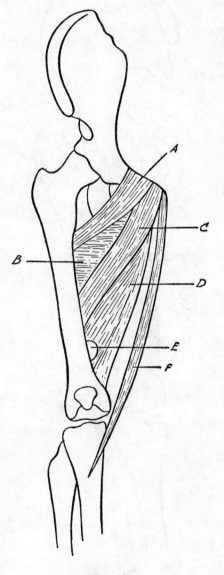

Fig. 276

the inferior part of the muscle meets the femur, there is a gap (E) in the muscle. This opening permits the passage of a large artery of the anterior aspect of the thigh, the femoral artery, to the posterior aspect of the knee joint. The **gracilis** (F), a slender muscle arising from the same general area as the other adductors, inserts on the tibia. These muscles all function to adduct the thigh.

The posterior surface of the adductor magnus muscle is illustrated in Fig. 277, which shows the posterior aspect of the pelvis and thigh. The broad adductor magnus consists of a transverse part (B) and a longitudinal, or **hamstring,** part

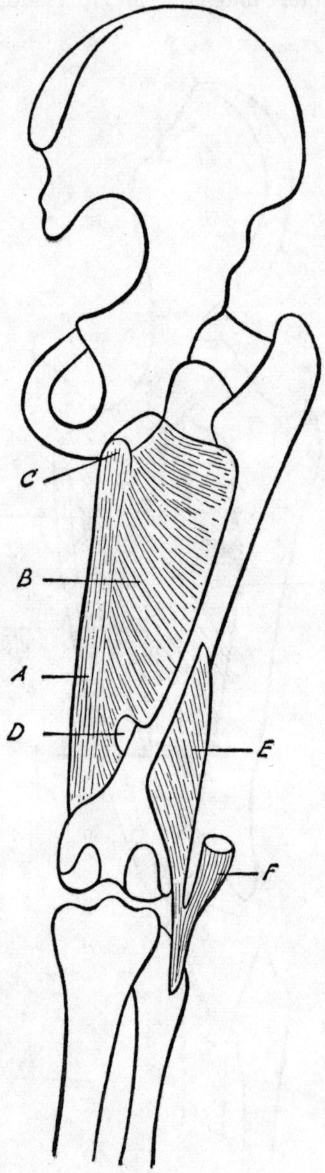

Fig. 277

(A). The hamstring arises from the ischial tuberosity (C) and inserts on the adductor tubercle. Because of the direction of these muscle fibers the hamstring can also extend the thigh. The opening in the muscle for the femoral artery (D) leads to the back of the knee joint. The deepest muscle of the flexor compartment (E) is the short head of the **biceps femoris,** which joins with its longer twin, the long head of the biceps (F), to form a common tendon that inserts on the head of the **fibula**. In Fig. 278 the remaining muscles of the flexor compartment are shown. They all arise from the ischial tuberosity. The broader **semimembranosus** (A) goes to the pos-

thigh the long head of the biceps femoris (C) joins with its short head (D) to insert on the head of the fibula.

THE KNEE JOINT

The muscles of the thigh have been followed to the tibia, or leg bone, where they either extend or flex it. These movements take place at the **knee joint.** The knee joint is the largest joint in the body because it is formed by two of the largest bones, the femur and tibia. Although the knee is subjected to considerable strain, it does not have the advantage of a deep bony socket, as is found in the hip joint. It must depend primarily on ligaments and muscles for its stability.

The general features of the joint are illustrated in Fig. 279. The rounded medial (A) and lateral

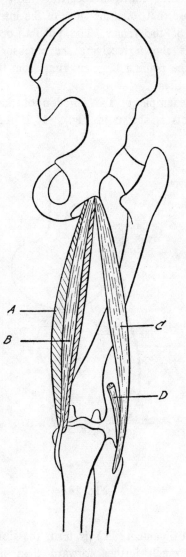

Fig. 278

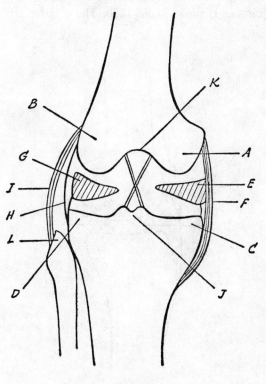

Fig. 279

terior aspect of the tibia, and the thinner **semitendinosus** (B) inserts on the antero-medial surface of the tibia in common with the **sartorius** and **gracilis** muscles. On the lateral side of the

(B) **condyles of the femur** are separated from the medial (C) and lateral (D) **condyles of the tibia** by two wedge-shaped cartilaginous discs called **menisci** (singular, meniscus). The **medial disc** (E) is firmly bound at its periphery to the fibrous capsule (F). The **lateral disc** (G) is more mobile and is not so closely bound to the capsule (H). The **lateral ligament** (I) of the knee joint is a separate cord, not a thickening of the capsule and is attached to the head of the fibula (L).

On the upper surface of the tibia between the condyles there are two little hills or tubercles joined by a plateau (J), which collectively form the **intercondylar eminence.** From this central region two cordlike ligaments arise and cross one another as they run to the **intercondylar notch** (K) of the femur. The two ligaments cross like the arms of an X and are therefore called the **cruciate ligaments.** They are the *internal* ligaments of the knee joint, whereas the lateral ligament (I) is an *external* ligament.

These cruciate ligaments give stability to the joint, as is shown in Fig. 280. The anterior cruci- ate ligament (A) arises anterior to the intercondylar eminence and passes upward and backward to be attached to the lateral condyle of the femur. The posterior cruciate (B) arises posterior to the eminence and runs upward and forward to the medial condyle. The surfaces of the femoral condyles receiving these ligaments are those that form the sides of the intercondylar notch of the femur. The **patella,** or kneecap (C), is shown receiving the extensor muscles, while the patellar ligament runs from the patella to the tibia.

An example of the function of the cruciate ligaments is shown in Fig. 281. If the anterior

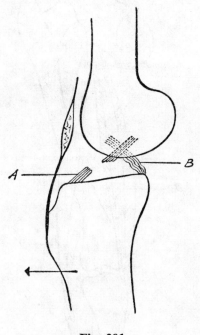

Fig. 281

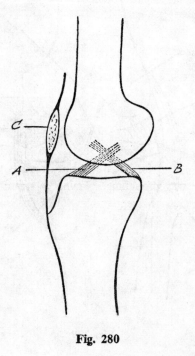

Fig. 280

cruciate ligament (A) is torn, the tibia can be pulled much farther forward than its normal movement. This forward movement of the tibia will not be hindered by the intact posterior cruciate ligament (B), which becomes slack. When the posterior cruciate is torn and the anterior cruciate remains intact, abnormal posterior movement of the tibia can be produced.

Menisci. The cartilaginous discs rest on the condyles of the tibia. In Fig. 282, the articular surface of the tibia is shown. The medial (A)

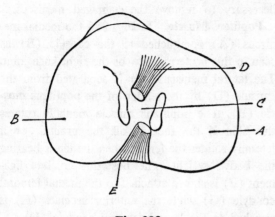

Fig. 282

and lateral (B) condyles of the tibia are separated by the intercondylar eminence (C). In front of the eminence a part of the anterior cruciate ligament (D) is seen. Posterior to the eminence is the cut portion of the posterior cruciate ligament (E). The cartilaginous discs, or menisci, are seen in Fig. 283. They resemble a

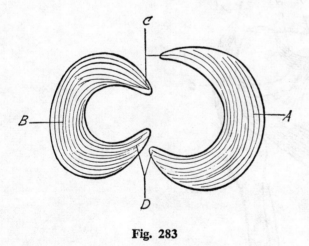

Fig. 283

half-moon in outline and are sometimes called the **semilunar cartilages.** The larger disc is the medial (A), but the lateral (B) is more circular.

The discs are attached to the tibia by their pointed ends, called **horns.** The anterior horns are shown at C, and the posterior horns at D. In Fig. 284 the medial (A) and lateral (B)

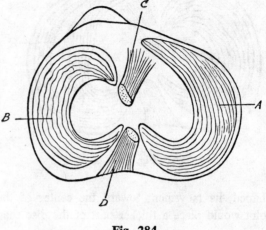

Fig. 284

cartilaginous discs are shown in their normal positions. The attachments of the horns are shown in relation to the anterior (C) and posterior (D) cruciate ligaments. The condyles of the femur are in contact with the discs, which act as shock absorbers during walking and jumping. There is movement not only between the femoral condyles and the menisci but also between the menisci and the condyles of the tibia during flexion and extension of the knee joint.

Athletes frequently injure a meniscus, the medial meniscus being most often affected. A typical history of such an injury reveals that the athlete was crouching, placing a strain on the medial side of the partially flexed knee. During a sudden movement in which the tibia was laterally rotated and abducted on the femur, he felt a sudden pain in the knee accompanied by a sensation of something going out of place. The knee joint was locked, that is, he could not extend it, and the limb was unable to bear any weight. Apparently the medial meniscus is drawn toward the center of the joint and is crushed by the medial femoral condyle during attempted extension of the joint. Since the meniscus is wedge-

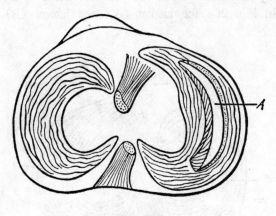

shaped, its movement toward the center of the joint would place a thicker part of the disc than

could be accommodated between the femoral and the tibial condyles. In Fig. 285 a typical tear of the medial meniscus is shown at A. Since a tear in fibrocartilage does not heal, it is often necessary to remove the damaged meniscus.

Popliteus Muscle. In Fig. 286 the medial meniscus (A) is attached to the capsule (B) as seen in this posterior view of the right knee joint. The lateral meniscus (C) is separated from the capsule (D) by the tendon of the **popliteus muscle** (E). The popliteus muscle medially rotates the tibia if the foot is off the ground, or it laterally rotates the femur when the leg is bearing the body's weight. The posterior cruciate ligament (F) is shown attached to the medial femoral condyle (G) while the anterior cruciate (H) is attached to the lateral femoral condyle (I).

On the lateral side of the knee joint, Fig. 287, the **lateral ligament** (A) passes from the lateral

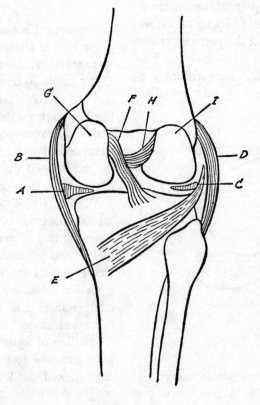

Fig. 286

epicondyle of the femur to the head of the fibula. The popliteus muscle tendon (B) separates the ligament from the lateral meniscus.

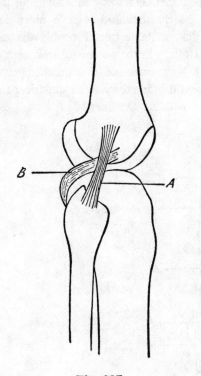

Fig. 287

Injuries. On the medial side of the knee joint, Fig. 288, the medial ligament (A) is a thickened part of the joint capsule, which is not otherwise shown here. This ligament is frequently injured during sudden tibial abduction strains, such as a forceful blow on the lateral side of the knee that might be received in a hard tackle in football. The patella (B) receives the insertion of the four extensor muscles (C). The patella is anchored to the tibia (D) by the patellar ligament (E). A badly fractured patella may be surgically removed without resulting in any major disability, provided that the extensor tendon (C) is sutured to the patellar ligament (E).

Synovial Cavity. The synovial cavity of the knee joint is related to the articular surfaces of the femur and tibia. In addition, the synovial cavity extends superiorly between the extensor

tendon (C) and the anterior surface of the femur. This extension of the synovial cavity is called the **suprapatellar bursa** (F), and may extend two inches above the superior border of the patella. Stab wounds in this area may therefore introduce infection into the synovial cavity. If excessive fluid collects in the synovial cavity as a result of injury or infection, the thigh swells just above the patella because of the fluid in the suprapatellar bursa.

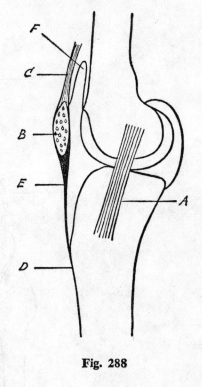

Fig. 288

Extensor Muscles. In Fig. 289 the four extensor muscles are seen inserting into the patella (F), while the patella is anchored to the tibia by the patellar ligament (G). This tendon of insertion is frequently referred to as the **quadriceps tendon** because it arises from four heads. The vastus lateralis (A), vastus medialis (B), vastus intermedius (C), and the cut portion of the rectus femoris (D) constitute the four heads. Arising from the vastus lateralis and vastus medialis are sheets of strong connective tissue (I and H) that pass distally over the knee joint to attach to the tibia. These sheets of tissue are called **patellar retinacula,** and form the anterior part of the capsule of the knee joint. Thus, the vasti muscles can directly affect the stability of the knee joint through their aponeurotic extensions, the patellar retinacula.

Injury to the knee joint initiates a cycle that leads to added insults to the joint. Following the initial injury, movement at the joint is decreased because of the pain; the quadriceps muscles then atrophy due to disuse. Atrophy of the quadriceps decreases their efficiency in stabilizing the joint, and a poorly stabilized joint is more easily injured than a stable one. Probably the most important factor in the successful treatment of knee injuries is exercise of the quadriceps muscles. How would you exercise the quadriceps muscles without bending the knee?

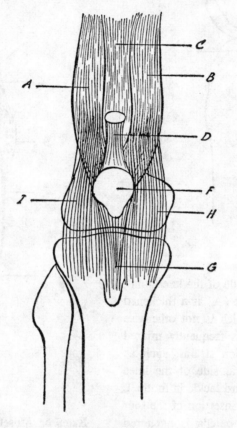

Fig. 289

Capsule. The components of the capsule of the knee joint are shown in Fig. 290. The femur has been removed, and you are looking down on the condyles of the tibia. The capsule is attached to the periphery of the condyles. On the medial side the capsule is thickened to form the **medial ligament** (A). The **lateral ligament** (B) is a cordlike structure separate from the capsule. The **patellar ligament** is shown at C. The fibrous sheets or patellar retinacula (D and E) form the anterior part of the capsule. The remaining portion of the capsule (F) is attached above to the femur and is similar to the fibrous capsule of other joints.

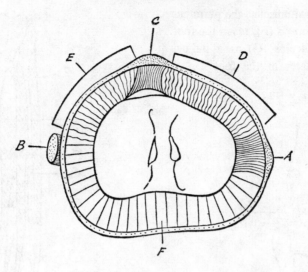

Fig. 290

THE LEG

The deep muscles of the posterior aspect of the leg are shown in Fig. 291. The **tibialis posterior**

Fig. 291

(A) is overlapped on the medial side by the **flexor digitorum longus** (D) and on the lateral side by the **flexor hallucis longus** (C). The tendons cross the ankle joint and reach the foot, where they will be shown later. The tibialis posterior tendon grooves the **medial malleolus** (B) and the other two pass under the **sustentaculum tali** (E).

There are two superficial muscles forming the calf of the leg, shown in Fig. 292, and together they are usually called the **triceps surae.** The larger and deeper muscle is the massive **soleus** (A), which arises from both fibula and tibia. The lateral head of the gastrocnemius (B) unites with its medial head (C) to form the **gastrocnemius muscle** (D), which unites with the soleus. From this union a common tendon (E), the famed **Achilles' tendon,** runs to insert on the heel bone or **calcaneus.** The triceps functions chiefly at the ankle joint to plantar flex the foot.

Three muscles on the anterior aspect and two muscles on the lateral aspect of the leg are illustrated in Fig. 293. The **tibialis anterior** (A) and the **extensor digitorum** (C) **muscles** partially cover the **extensor hallucis longus** (D). The tendon of the tibialis anterior muscle inserts on the foot bones (B), while the other tendons insert on the toes. The two lateral muscles, the **peroneus longus** (E) and **peroneus brevis** (F), reach the foot by using the **lateral malleolus** (G) as a pulley. They will be followed later in the foot.

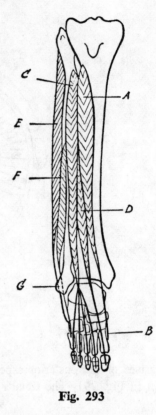

Fig. 293

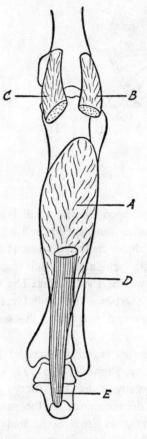

Fig. 292

NERVES AND VESSELS OF THE LOWER LIMB

The major artery of the lower limb, Fig. 294, is the **femoral artery,** which is a continuation of the external iliac artery (A) of the pelvis. At the root of the limb the external iliac artery becomes the femoral artery (B). The femoral ar-

tery soon gives off a major branch, the **profunda femoris artery** (C), which then divides into a **lateral femoral circumflex** (D) and a **medial femoral circumflex artery** (E). The lateral femoral circumflex divides into an ascending branch going to the hip and a descending branch going to the knee. The medial femoral circumflex artery passes posteriorly to supply the hip joint. The profunda femoris passes distally to supply the deep muscles of the thigh. The femoral artery continues distally to pass through the opening (F) in the adductor magnus muscle to reach the back of the knee joint. Another artery emerging through the obturator foramen from the pelvis, the **obturator artery,** sends one branch (G) to the hip joint, and the other branch (H) supplies the adductor muscles. The nerve (J) adjacent to the obturator artery is the **obturator nerve,** the chief motor nerve supply to the adductor muscles. The nerve (I) lying lateral to the femoral artery is the **femoral nerve.** It supplies the quadriceps muscles and the sartorius muscle. Other branches supply the skin on the anterior and medial surfaces of the thigh.

Fig. 294

In the extensor compartment of the thigh, Fig. 295, the huge **sciatic nerve** arises in the pelvis and passes through the greater sciatic foramen (A) to lie on the ischium (B), where it may be injured in posterior dislocations of the hip joint. It passes midway between the ischial tuberosity and the lesser trochanter (C). It gives muscular branches (D) to all the extensor muscles, including the hamstring part of the adduc-

tor magnus muscle. Above the knee joint (E) the sciatic nerve divides into a lateral or **common peroneal branch** (F) and a medial or tibial branch (G). The common peroneal nerve passes around the posterior surface of the head of the fibula, where you can feel it by rolling it under

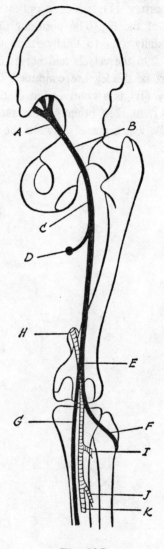

Fig. 295

your finger against the bone. It continues around the fibula to reach the anterior surface of the leg. The **tibial nerve** (G) supplies all the flexor muscles in the posterior compartment of the leg.

As soon as the femoral artery emerges through the opening of the adductor magnus muscle (H), it is called the **popliteal artery.** After giving off the anterior tibial artery (I), which passes between the tibia and fibula to the front of the leg, the popliteal becomes the **posterior tibial artery.** The posterior tibial artery gives rise to the **peroneal artery** (J), which supplies muscles in the back of the leg. The posterior tibial artery passes distally (K) to finally reach the foot.

In Fig. 296 the vessels and nerves in the posterior part of the leg are continued. The popliteal artery (B) is a continuation of the femoral artery (A). Its chief branch is the anterior tibial artery (C), which passes between the leg bones.

The posterior tibial artery (D), the continuation of the popliteal artery, gives off the peroneal artery (E), which passes distally above the ankle joint where it passes between the leg bones to the anterior part of the ankle. The posterior tibial artery (F) continues distally to cross the ankle joint (L) to reach the foot.

The sciatic nerve (G) divides into the common peroneal nerve (H) and the tibial nerve (I). The tibial nerve gives branches to the triceps surae (J) and to the tibialis posterior (K), flexor digitorum longus, and flexor hallucis longus muscles. It continues distally to accompany the posterior tibial artery to the foot.

In the anterior part of the leg, Fig. 297, the

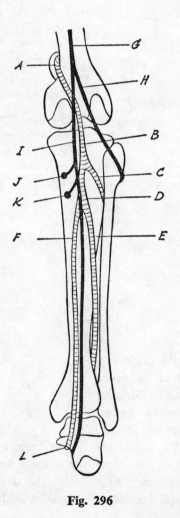

Fig. 296

anterior tibial artery (E) is seen emerging from between the tibia and fibula to pass distally to the ankle joint. When it crosses the ankle joint the anterior tibial artery is called the **dorsalis pedis artery** (F) and runs forward on the dorsal surface of the foot. The distal end of the peroneal artery (G) emerges between the leg bones and runs on the dorsal surface of the foot to form an arterial arch (H) with the dorsalis pedis artery. From this arch, small branches pass to the toes.

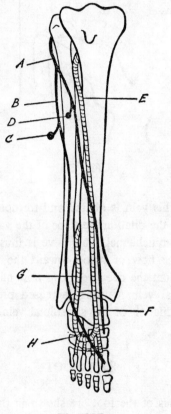

Fig. 297

The common peroneal nerve (A) appears on the lateral aspect of the fibula, where it soon divides into a superficial peroneal branch (B) and a deep peroneal branch. The superficial peroneal nerve supplies the peroneus longus and brevis muscles (C) and continues distally in the leg and foot, where it supplies skin, chiefly. The deep peroneal branch (D) supplies the tibialis anterior, extensor hallucis longus, and extensor digitorum longus muscles.

The veins of the lower limb for the most part follow the same paths as the arteries and have the same names as the arteries they accompany. Since the arteries are found between muscles, the veins are also deep, although in addition to the deep veins there is a system of superficial veins that are unaccompanied by arteries and lie just under the skin. These veins drain blood from superficial structures, such as skin, and convey the blood to the deep veins. A diagram of the general plan of superficial and deep veins is shown in Fig. 298. The deep vein is situated

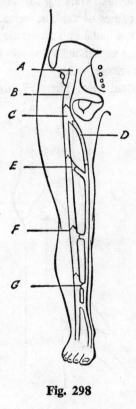

Fig. 298

in the center of the limb, draining blood from the foot, leg, and thigh. The femoral vein (B) becomes the external iliac vein (A) in the pelvis. The superficial vein (D) lying adjacent to the skin drains the superficial structures of the foot, leg, and thigh and enters the femoral vein at A. This superficial vein is known as the long saphenous vein. In addition to the main junction (C) it has numerous communicating channels (E and G). Both superficial and deep veins have valves (F) that allow the blood to flow in one direction—toward the heart.

The long saphenous vein frequently becomes **varicose,** that is, dilated and tortuous. When a person with varicose veins stands for a short period of time the dilated veins fill with blood and thus are very prominent. The causes of varicose veins are not clearly understood, but it appears that any increase in the pressure of the blood in the superficial veins will aggravate the condition. Pregnancy and standing for long periods are contributing causes of varicose veins. Conversely, elevation of the limbs and an elastic support that compresses the superficial veins will ease the discomfort of varicose veins.

In the normal vein, Fig. 299, the deep vein

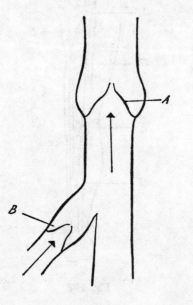

Fig. 299

has a valve (A) that usually consists of two flaps, which meet in the middle of the channel and allow blood to flow in the direction of the

arrow. The valve in the superficial vein (B) also allows blood to flow only into the deep vein. In Fig. 300 the superficial vein (C) is vari-

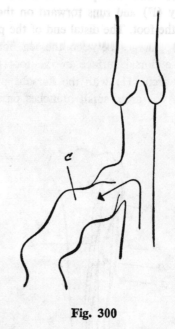

Fig. 300

cosed. This vein is dilated and tortuous, and because of the dilation the flaps of the valve cannot meet in midchannel. The valve is thus unable to direct the flow of blood toward the deep vein. Blood from the deep vein can now pass into the superficial vein, causing increased pressure and further dilation of the superficial vein.

THE FOOT

The bones of the foot are shown in three views: from the medial side of the foot, Fig. 301, from the lateral side of the foot, Fig. 302, and as seen from the plantar aspect or the sole of the foot, Fig. 303. The bones of the foot are called **tarsal** and **metatarsal** and correspond to the carpal and metacarpal bones of the hand. The **talus** (A) articulates with the distal ends of the tibia and fibula to form the ankle joint. The weight of the body is transmitted from the tibia to the talus and then to the remaining tarsal bones. The largest tarsal bone is the **calcaneus** (B). The heel of the foot is formed by the posterior end of the calcaneus. Anterior to the talus is the **navicular bone** (C). Anterior to the navicular are three wedge-shaped tarsal bones called **cuneiform bones;** they are named according to their relative positions to each other—the **medial** (D),

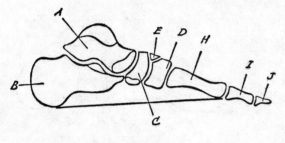

Fig. 301

the **intermediate** (E), and the **lateral** (F) cuneiform bones. Anterior to the calcaneus is the **cuboid bone** (G). The five metatarsal bones are indicated by H. The phalanges of the toes are similar to those of the hand. The great toe, like the thumb, has only two phalanges, and all the other toes have three, the **proximal** (I), **middle,** and **distal** (J) phalanges.

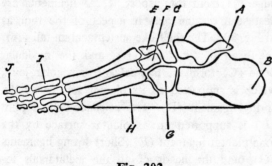

Fig. 302

Probably more important than the individual bones is the architecture of the foot bones. The tarsal and metatarsal bones form a series of longitudinal arches. One end of the arch is formed by the calcaneus and the other end by the heads of the metatarsals. The keystone of the arch is the talus. The height of the longitudinal arch on the medial side of the foot is greater than that on the lateral side. If you examine the print of a bare foot, you can follow the outline of the lateral border of the foot, but the outline of the medial side is not complete. The medial arch

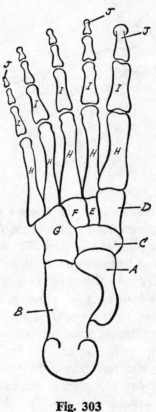

Fig. 303

is normally not sufficiently depressed by the body weight to make contact with the ground.

Although a solid bony arch is a most efficient method for supporting weight, an arch composed of movable bones also provides a mechanism that absorbs shocks and can automatically adjust when walking over uneven ground.

Foot disorders are exceedingly common, and you frequently hear the term "flat feet," which means depression of the medial longitudinal

arches. While some people have painful flat feet,
not all flat feet are painful. We will examine
the joints between the tarsal bones and their supporting structures in order to understand how the
longitudinal arch may be depressed.

Medial Longitudinal Arch. The bones forming the medial longitudinal arch are shown in
Fig. 304. The body of the talus and the calcaneus

this ligament is apparent, for in depression of
the medial longitudinal arch the head of the talus
is forced nearer to the ground. One structure
limiting this inferior movement of the head of
the talus is the spring ligament.

Lateral Longitudinal Arch. The bones making up the lateral longitudinal arch are shown
in Fig. 305. The talo-calcaneal joint is indicated

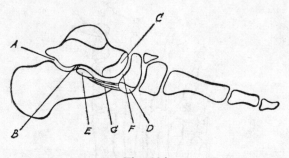

Fig. 304

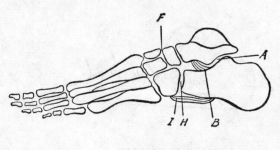

Fig. 305

articulate (A) to form the **talo-calcaneal joint.**
During inversion and eversion of the foot, movement takes place at this joint. The **interosseous
talo-calcaneal ligament** (B) holds the two bones
together. Try to visualize how the ligament would
prevent the talus from being pushed forward on
the calcaneus by the weight of the body when
you are standing on tiptoe or even walking with
high-heeled shoes. The head of the talus (C)
articulates with two bones, the navicular (D)
and a shelf-like projection of the calcaneus (E)
called the **sustentaculum tali.** This shelf projects
medially, that is, toward you, and it supports the
inferior part of the head of the talus. Since this
joint is formed by three bones, it is called the
talo-calcaneo-navicular joint (F). Notice the
bony gap between the sustentaculum tali and
the navicular bone. A strong ligament stretching
between these two bones provides an important
support of the head of the talus. The ligament
is on the plantar aspect of the foot, so in this
diagram you are looking at its medial edge (G).
This ligament is called the **plantar calcaneo-
navicular,** or **spring ligament.** The importance of

by A, the interosseous talo-calcaneal ligament by
B. The talo-calcaneo-navicular joint is partially
visible at F. An important joint is formed between the calcaneus and the cuboid (H) and,
as you might expect, is called the **calcaneo-cuboid joint.** Since the longitudinal arch may be
depressed at this joint, a strong ligament is found
on its plantar aspect. This ligament is called the
long plantar ligament, and you can see its lateral
edge (I) from this aspect. These ligaments are
best seen on the plantar aspect of the foot as
in Fig. 306. The shelf-like sustentaculum tali (A),
the head of the talus (B), and the navicular
bone (C) form the talo-calcaneo-navicular joint,
which is supported on its plantar surface by the
spring ligament (D). The calcaneo-cuboid joint
(E) is supported on its plantar surface by the
long plantar ligament (F). Short strong ligaments
(G) bind the heads of all the metatarsals together and are called the **deep transverse metatarsal ligaments.** It should be noted that the
head of the first metatarsal is bound to that of
the second, which precludes any grasping movements of the big toe. You may remember that
in the hand the head of the first metacarpal is

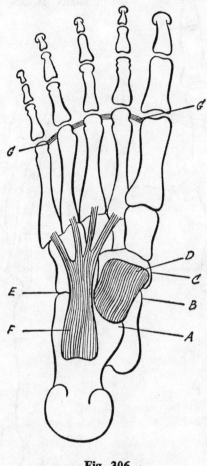

Fig. 306

tendons and muscles of the foot is shown in Fig. 307. The peroneus longus tendon (A) arises from the lateral side of the leg and bends around the lateral border of the foot. It is partially under cover of the long plantar ligament (B) and inserts on the base of the first metatarsal and adjacent medial cuneiform. This tendon everts and plantar flexes the foot. The tibialis anterior tendon (C) inserts on the opposite side of the same two bones. This muscle inverts and dorsi-

flexes the foot because the tendon passes anterior to the ankle joint. The tibialis posterior tendon (D) divides into slips that insert on several of the tarsal bones. The tibialis posterior muscle, arising from the posterior aspect of the leg, uses the medial malleolus as a pulley to plantar flex and invert the foot. The interossei muscles in the foot have the same general plan as in the hand. Since the toes have little independent movement, they act together to flex the metatarso phalangeal joints. They probably prevent displacement of the metatarsal bones during weight bearing. The three plantar in-

muscle (D). The names of the muscles indicate their function. A small sesamoid bone is found in each tendon of the flexor hallucis brevis just proximal to its insertion. The flexor digiti minimi brevis (F) arises from the base of the fifth metatarsal and inserts on the base of the proximal phalanx. It aids in flexing the little toe.

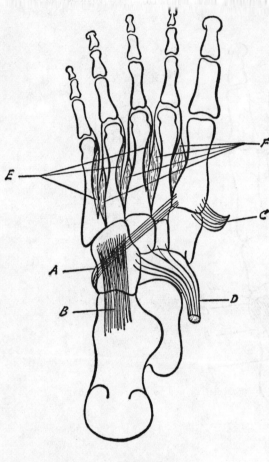

Fig. 307

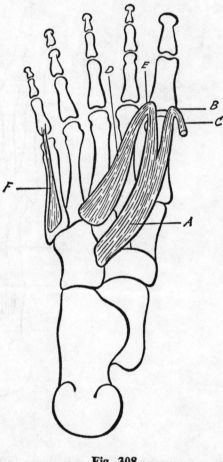

Fig. 308

terossei muscles are indicated by (E) and the four dorsal ones by (F).

Fig. 308 shows a group of three muscles that lie just superficial to the structures in the preceding figure. The flexor hallucis brevis (A) arises from the tarsal bones and divides into two tendons. The medial tendon inserts on the base of the proximal phalanx (B) in common with the tendon of the abductor hallucis (C). The lateral tendon inserts on the proximal phalanx (E) with the tendon of the adductor hallucis

Lying immediately superficial to the preceding structures are the long flexor tendons of the foot, Fig. 309. The flexor hallucis longus tendon (A) passes inferior to the sustentaculum tali and the spring ligament (B) to insert on the base of the distal phalanx of the big toe. It reinforces the spring ligament and thus aids in maintaining the longus tendon (C) takes a more oblique course, and a slip from the tendon is inserted on the

(D) on

digiti minimi arises from the calcaneus and inserts on the lateral side of the base of the proximal phalanx. The flexor digitorum brevis gives rise to four tendons, each of which splits just before inserting on the base of the middle phalanx (C) and assists the flexor digitorum longus

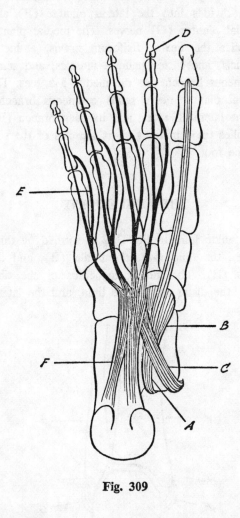

Fig. 309

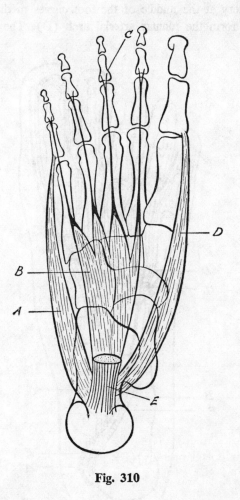

Fig. 310

distal phalanx of each toe (D). As in the hand, a small wormlike muscle arises from each slip (E). These are the four **lumbrical muscles,** which are inserted on the extensor tendons of the lateral four toes. The **accessory flexor** (F) is inserted into the long flexor tendon, correcting for the oblique pull of the flexor digitorum longus tendons on the toes. By pulling on the long flexor tendons, the accessory flexor converts the oblique pull to a more direct posterior pull.

The most superficial group of muscles, Fig. 310, is made up of the **abductor digiti minimi**

muscle. Superficial to the flexor digitorum brevis is a thickened layer of fascia (E) called the plantar aponeurosis, which runs toward the toes to attach to the fibrous sheaths surrounding the tendons in the toes. It is similar to the palmar aponeurosis in the hand. The plantar aponeuroalo, although cut in this material, is a strong, fibrous support for the longitudinal arch of the foot.

Nerves and Arteries. The nerves and arteries of the sole of the foot are shown in Fig. 311. The **posterior tibial artery** (A) divides to form the **lateral plantar artery** (B) and the **medial plantar artery** (C). The medial plantar artery continues toward the big toe. The lateral plantar artery at the middle of the foot curves medially to form the **plantar arterial arch** (D). The ar-

appear on the sole, interosseal branches arising from the arterial arch are seen passing to the toes. The **posterior tibial nerve** (E), like the artery, divides into the lateral plantar (F) and medial plantar (G) nerves. The medial plantar supplies the flexor digitorum brevis, abductor hallucis, and flexor hallucis muscles, and sends cutaneous branches to the medial 3½ toes. The lateral plantar nerve sends cutaneous branches to the lateral 1½ toes, and its deep branch (H) supplies the remaining short muscles of the sole of the foot.

THE ANKLE JOINT

The ankle joint in Fig. 312 is formed by three bones, the tibia (A), the fibula (B), and the talus (E). The medial malleolus of the tibia (C), the distal end of the tibia, and the lateral

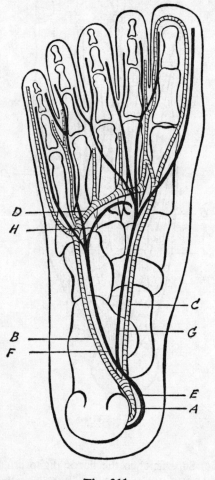

Fig. 311

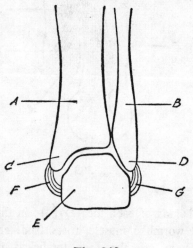

Fig. 312

terial arch continues to the space between the first and second metatarsals, where it anastomoses with the dorsalis pedis artery from the dorsal

malleolus (D) of the fibula form a rigid mold into which the body of the talus snugly fits. Movement in only one plane, that of plantar and dorsiflexion of the foot, is permitted. The body malleoli prevent abduction and adduction. Strong ligaments on the medial (F) and lateral (G) sides of the ankle joint guard against such movements.

ankle injuries. When forcible side-to-side movement occurs, such as suddenly "going over" on the ankle, these ligamentous structures are frequently torn. On the medial side of the ankle joint, Fig. 313, the ligament is called the **deltoid.** It is attached to the tip of the medial malleolus (A) and runs to the body of the talus (B), to the sustentaculum tali (C), and to the navicular bone (D). It is injured in forcible eversion of the foot. The ligament is so strong that it usually tears off the bony tip of the malleolus.

On the lateral side of the joint, Fig. 314, three protective ligaments are found. They are all attached to the tip of the lateral malleolus (A).

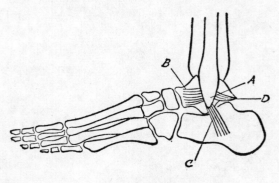

Fig. 314

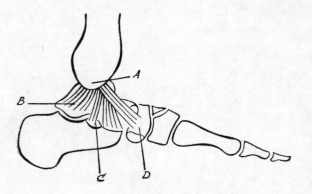

Fig. 313

The **anterior talo-fibular** (B), the **calcaneo-fibular** (C), and the **posterior talo-fibular** (D) ligaments comprise the lateral ligaments of the ankle joint. They are frequently torn when the foot is forcibly inverted, and the first two (B and C) are usually the victims.

INDEX

Liver, 92
Lumen, 34
Lungs, 78–80
Lymphatic system, 36
Lymph nodes, 36
Lymphocytes, 36

Macula, 60
Macula lutea, 52
Malleus, 55, 56
Mandible, 37, 39
Marrow, 19–20, 37
Mastication, 65–66
Mastoid bone, 37
Maxillary bone, 37, 62
Meatus, 61
Median plane, 13
Median septum, 60, 62
Medulla, 44
Membrane bones, 37
Membranes: bones, 37; cavital, 75–76; cranial, 38, 40, 41; nasal (mucous), 64; of anal canal, 107; of heart, 80; sacs in ear, 58; synovial, 21, 146; tympanic, 54; urethra, 96
Meninges, 46
Menisci, 165–66
Midbrain, 44
Middle ear, 54, 57
Mitral value, 82
Motor nerves, 25, 28, 29
Mucus, 61
Muscles, 22–24; involuntary, 23; neck, 71, 73, 74; of abdominal wall, 84–86; of the arm, 125–27; of the breast, 128; of ear, 55; of eye, 49–50, 53; of the foot, 177–80; of the forearm, 133–36; of the gluteal region, 155–56; of the hand, 142–46; of knee joint, 166–68; of lower leg, 169–70; of mastication, 65; of the penis, 101; of the shoulder, 123–25; of thigh, 158–63; of tongue, 67–68; pharynx, 69; voluntary, 22–23
Myelin, 26

Nasal bone, 39
Nasal cavity, 61
Nasopharynx, 55
Neck, 71–74
Necrosis, 35
Nerves: cells of, 25–26; cranial, 27–28, 47–48; destruction of, 27; facial, 65; in muscles, 23, 155; of foot, 180; of hand, 146; of limb muscles, 138–40; of lower limb, 170–73; reflexes, 29; spinal, 28; sympathetic, 30–32
Neuron, 25

Neutrophils, 19
Nodes of Ranvier, 26
Nose, 60–64; blood supply of, 63–64; bones of, 62–63; mucous membranes, 64; nasal cavity, 61–62
Nostril, 60
Nucleus, 25

Occipital bone, 37, 39
Oculomotor nerve, 47
Olfactory bulb, 61
Olfactory nerve, 47
Optic nerve, 47, 49
Orbicularis oculi muscle, 53
Orbit, 49
Organ of Corti, 58
Ossicles, 55
Ossification centers, 18
Otitis media, 57
Otoliths, 60
Otosclerosis, 57
Ovary, 103, 105
Oxygen-carbon dioxide exchange, 33

Palatine bone, 62
Pancreas, 87–88
Parasympathetic nervous system, 30, 32
Parietal bone, 37, 39
Parotid gland, 65
Patella, 159, 164
Pecten, 107
Pectoralis major, 126, 128
Pelvic cavity, 75, 93–96
Pelvic diaphragm, 108–10
Pelvic girdle, 150–52
Pelvis, female, 110–12
Pelvis, male, 110
Penis, 96, 98
Perilymph, 58
Periosteum, 20, 21
Peripheral nervous system, 26
Peritoneum, 76, 92–93
Petrous bone, 37
Pharynx, 60, 61, 68–71
Pineal gland, 44
Pituitary gland, 41
Plasma, 33
Pleura, 76
Pleural cavity, 78
Pleurisy, 78
Position: anatomical, 13; of structures to each other, 14–16
Processus vaginalis, 100
Prostate gland, 95, 96–97